Teaching Statistics and Probability

1981 Yearbook

Albert P. Shulte
1981 Yearbook Editor
Oakland Schools, Pontiac, Michigan

James R. Smart
General Yearbook Editor
San Jose State University

National Council of
Teachers of Mathematics

Library of Congress Catalog Card Number: 81-1679
ISBN 0-87353-170-1

Printed in the United States of America

Table of Contents

Preface ix

Part I: The Case for Teaching Statistics and Probability

1. Why Teach Statistics and Probability—a Rationale 1

    Lionel Pereira-Mendoza, Memorial University of Newfound-
    land, St. John's, Newfoundland
    Jim Swift, Nanaimo District Senior Secondary School, Na-
    naimo, British Columbia

    An introduction to the other articles in the yearbook, demonstrating
    that statistics and probability should be taught for utility, for future
    study, and for aesthetic reasons.

Part II: Samples of Existing Courses or Programs

2. Experiential Statistics and Probability for
Elementary Teachers 8

    William A. Juraschek, University of Colorado at Denver, Den-
    ver, Colorado
    Nancy S. Angle, University of Colorado at Denver, Denver,
    Colorado

    An activity approach to statistics and probability for prospective ele-
    mentary teachers.

3. Teaching Statistics to Eleven-to-Sixteen-Year-Olds: An
Account of Some Development Work in England and Wales . . . 18

    Peter Holmes, Schools Council Project on Statistical Education,
    University of Sheffield, Sheffield, England
    Daphne Turner, Schools Council Project on Statistical Educa-
    tion, University of Sheffield, Sheffield, England

    A widely tested project using topical modules to teach statistics to
    British students.

4. An Elementary Course in Nonparametric Statistics 24

Albert M. Liebetrau, Johns Hopkins University, Baltimore, Maryland

Nonparametric statistics is increasingly important. A sample course is outlined, and some detailed examples are presented.

Part III: Classroom Activities

5. Graphically Speaking: Primary-Level Graphing Experiences 33

Laura Duncan Choate, La Paloma School, Fallbrook, California
JoAnn King Okey, La Paloma School, Fallbrook, California

(Preschool–Grade 3)   A range of graphing activities, moving from concrete to abstract.

6. Using Teaching Devices for Statistics and Probability
   with Primary Children 41

Kathleen A. Knowler, Arizona State University, Tempe, Arizona
Lloyd A. Knowler, University of Iowa, Iowa City, Iowa

(Kindergarten–Grade 3)   Primary-level activities—graphing, sorting, measuring, estimating, playing games—using statistics and probability.

7. Statistical Sampling and Popsicle Sticks 45

Jerry Shryock, Western Illinois University, Macomb, Illinois

(Grades 2–6)   Sampling from a population of colored Popsicle sticks and making predictions about the population.

8. Fair Games, Unfair Games 49

George W. Bright, Northern Illinois University, DeKalb, Illinois
John G. Harvey, University of Wisconsin—Madison, Madison, Wisconsin
Margariete Montague Wheeler, Northern Illinois University, DeKalb, Illinois

(Grades 4–8)   Appealing classroom games to refine the notions of fairness and unfairness.

9. Developing Some Statistical Concepts in the
   Elementary School 60

Harry Bohan, Sam Houston State University, Huntsville, Texas
Edith J. Moreland, Houston Independent School District, Houston, Texas

(Grades 3–6)   Graphical and pictorial techniques for determining the mean, median, and mode.

10. Triples: A Game to Introduce the Facts of Chance 64

Eris Bailey, Towson Senior High School, Towson, Maryland

(Grades 6–8)   Using the game of triples to reinforce computation, to introduce probability, and to practice decision making.

11. Random Digits and Simulation 70

> *Edward Silver,* San Diego State University, San Diego, California
>
> *J. Philip Smith,* Southern Connecticut State College, New Haven, Connecticut
>
>> (Grades 6-9)   Generating random digits and examining sequences for randomness. Simulating a water-flow situation.

12. The Student Price Index 73

> *C. Gail Tibbo Lenoski,* Statistics Canada, Vancouver, British Columbia
>
>> (Grades 10 12)   Making a student price index as a class project.

13. The Statistics Odds Room 83

> *Thomas E. Obremski,* Ohio State University, Columbus, Ohio
>
>> (Grades 10-12)   Unusual and motivating games of chance used at Ohio State's annual High School Day.

Part IV: Teaching and Learning Specific Topics

14. Misconceptions of Probability: From Systematic Errors to Systematic Experiments and Decisions 90

> *J. Michael Shaughnessy,* Oregon State University, Corvallis, Oregon
>
>> Common misconceptions about probability. Decisions based on the misconceptions *representativeness* and *availability.*

15. Some Statistical Paradoxes 100

> *H. E. Reinhardt,* University of Montana, Missoula, Montana
>
>> A collection of paradoxes and apparent paradoxes. Useful for class exploration and discussion.

16. Simple Graphical Techniques for Examining Data Generated by Classroom Activities 109

> *Carolyn Alexander Maher,* Rutgers University, New Brunswick, New Jersey
>
>> Applying new techniques (the stem-and-leaf plot and the median-hinge box-plot) to random walks and Markov chains.

17. Random Digits and the Programmable Calculator 118

> *Lennart Råde,* Comprehensive School Mathematics Program, St. Louis, Missouri (Gothenburg, Sweden)
>
>> How to generate random digits with a programmable calculator. Applying random digits in simulation activities.

18. Correlation, Junior Varsity Style 126

Annette N. Matsumoto, University of Hawaii Laboratory
School, Honolulu, Hawaii

A presentation of correlation at a level appropriate for junior high
school.

19. An Area Model for Solving Probability Problems 135

Richard D. Armstrong, Comprehensive School Mathematics
Program, St. Louis, Missouri

Applying area notions to solving probability problems. Appropriate
for upper elementary grades through junior high school.

20. Geometrical Probability 143

Richard Dahlke, University of Michigan—Dearborn, Dearborn,
Michigan
Robert Fakler, University of Michigan—Dearborn, Dearborn,
Michigan

A thorough development of geometric probability with examples and
applications appropriate for high school or community college.

21. Evaluating Exact Binomial and Poisson Probabilities
without Tables 154

Harry O. Posten, University of Connecticut, Storrs, Connecticut

A computer algorithm for generating classical probability distributions.

Part V: Applications

22. Paradoxes in Sampling 162

Clifford H. Wagner, Pennsylvania State University—Capitol
Campus, Middletown, Pennsylvania

Use and misuse of the principle of indifference. Why the famous
Literary Digest poll of 1936 was wrong. Present-day applications for
sampling.

23. Using *Consumer Reports* as a Resource for Data Analysis
in the Statistics Classroom 168

John W. McConnell, Glenbrook South High School, Glenview,
Illinois

Choosing stereo loudspeakers and evaluating the quality of food at
fast-food chains.

24. Applications of Statistics and Probability to Genetics 173

Hunter Ballew, University of North Carolina, Chapel Hill,
North Carolina

Looking at the role of statistics and probability in the development of
genetics from Gregor Mendel's time to the present day.

Part VI: Statistical Inference

25. Activities in Inferential Statistics 182

Barry V. Kissane, University of Western Australia, Nedlands, Australia

Making inferences based on sampling a population. The central limit theorem and its role in statistical inference.

26. Statistical Inference in Junior High and Middle School 194

Walter J. Sanders, Indiana State University, Terre Haute, Indiana

Studying chance variability in sampling. Using sampling results to infer population characteristics.

Part VII: Monte Carlo Techniques and Simulation

27. Monte Carlo Simulation: Probability the Easy Way 203

Ann E. Watkins, Los Angeles Pierce College, Los Angeles, California

How to use Monte Carlo simulations in the classroom. A number of motivational examples.

28. Using Monte Carlo Methods to Teach Probability and Statistics 210

Kenneth J. Travers, University of Illinois at Urbana, Urbana, Illinois

A brief history of Monte Carlo methods. Examples applying these methods to a variety of areas in statistics and probability.

Part VIII: Using Computers

29. Solving Probability Problems through Computer Simulation ... 220

William Inhelder, Herbert Hoover High School, Glendale, California

A computer program for investigating extrasensory perception.

30. In All Probability, a Microcomputer 225

Howard M. Kellogg, Fiorello H. LaGuardia Community College, Long Island City, New York

Using a microcomputer to study a tontine, to estimate the area of a region by Monte Carlo techniques, and to simulate the Buffon needle problem.

Bibliography (compiled by *Stuart A. Choate*) 234

Projects (compiled by *Ralph H. Klitz, Jr.*) 239

Preface

Statistics and probability are appropriate topics in the school mathematics curriculum because (1) they provide meaningful applications of mathematics at all levels; (2) they provide methods for dealing with uncertainty; (3) they give us some understanding of the statistical arguments, good and bad, with which we are continually bombarded; (4) they help consumers distinguish sound use of statistical procedures from unsound or deceptive uses; and (5) they are inherently interesting, exciting, and motivating topics for most students.

Recognizing the importance and appropriateness of these topics, the Publications Committee of the NCTM has selected statistics and probability as the theme of this 1981 Yearbook. As the list of authors in the Table of Contents indicates, a dedicated body of individuals all over the world are committed to expanding the teaching of statistics and probability in the schools. All major curriculum groups in this century—including the NCTM in its recommendations for the curriculum of the 1980s—have stressed the importance of statistics and probability. In spite of this, relatively little instructional time is given to these topics in most school systems, and relatively little material is available to offer teachers suitable classroom ideas. We hope the material in this yearbook will capture your interest and give you a springboard for beginning the teaching of statistics and probability. For those of you who have already discovered the excitement of teaching these topics, we hope it will give you new inspiration and ideas.

The first article ties together all the other articles in the yearbook and makes the case for teaching statistics and probability. The next several articles discuss units and courses in U.S. schools and a project in statistical education carried out in England and Wales. A distinctive feature of this volume is the inclusion of a variety of actual classroom-tested activities spanning all grades, K–14. These articles are an attempt to increase the usefulness of the material to the classroom teacher. To aid teachers in broadening their knowledge of statistics and probability, a number of specific topics are treated in essay form, including such new ideas as graphical data analysis and nonparametric statistics.

Some articles show applications of statistics and probability in the world outside the classroom. The major branches of statistics—descriptive statistics and inferential statistics—each receive major attention. The simulation of real-life situations is approached through the powerful Monte Carlo method. Suggestions are made for using computers, microcomputers, and programmable calculators in connection with the study of statistics and probability to enhance the learning that can take place.

Seventy-eight articles were submitted and considered for this yearbook. The task of reviewing the articles, selecting those to be included, and making suggestions for the revision of those selected was carried out by the editors and a dedicated and knowledgeable advisory panel. This panel read each manuscript at each stage of submission. Our thanks and appreciation go to the panelists: Stuart Choate (Midland, Michigan); Ralph Klitz, Jr. (Cincinnati, Ohio); Richard Pieters (Providence, Rhode Island); and Jim Swift (Nanaimo, British Columbia). Each of these panelists is a classroom teacher.

The editors further wish to thank the NCTM Publications Committee and the NCTM headquarters staff for their interest in this topic and their guidance through the process. Of course, we also wish to thank all those who submitted manuscripts for consideration.

It has been a pleasure to prepare this yearbook for you, the reader. If you have taught statistics or probability in the past, we believe the yearbook will offer useful new activities, techniques, and background material. If you have not taught these subjects, we hope you will be sufficiently excited about some of the ideas to try them out with your students.

ALBERT P. SHULTE
1981 Yearbook Editor

JAMES R. SMART
General Yearbook Editor

Reference Symbols

All bibliographical references are given in one listing at the end of the yearbook. The reader is directed to an entry in this listing by its corresponding number cited in the text in brackets.

I. The Case for Teaching Statistics and Probability

1. Why Teach Statistics and Probability—a Rationale

Lionel Pereira-Mendoza

Jim Swift

Even a cursory glance at newspapers shows the extent to which the language of statistics and probability has become a part of everyday life. Understanding this language has clearly become important. Articles 2, 3,

and 4 in this yearbook describe a variety of efforts designed to increase emphasis in the curriculum on the ideas of probability and statistics.

Even though the role of both statistics and probability in our lives is significant, it is not the only rationale for including them in the school curriculum. A model of a more complete rationale contains three components—utility, future study, and aesthetics.

 Individuals need a knowledge of statistics and probability to function in our society. Such things as consumer reports (see article 23), cost of living indexes (article 12), and surveys and samples (articles 7, 12, 22, 25, 26) are a part of everyday life. Students should be able to interpret such statements as "The probability of an oil spill off Vancouver Island is less than 1 in 10 000" or "Survey Spells Economic Gloom." Competence with the utilitarian aspects of statistics and probability will help them process the many data-oriented messages they receive every day.

A knowledge of statistics and probability is needed to deal with many situations that the student may face later in mathematics and other subjects. Competence in statistics and probability gives the student a sound basis for subjects that require this kind of mathematical orientation and foundation. Such a basis will become increasingly important as more fields of study require mathematical training. Subjects such as biology (article 24) and the social sciences (article 22), which at one time required little mathematical knowledge, are rapidly becoming dependent on sophisticated mathematical techniques, most of them statistical in nature. Simulation techniques and Monte Carlo methods (articles 27, 28) are now used in a wide range of disciplines. In a rapidly changing world, the assessment of probabilities of future events is an important part of decision making (article 20).

Aesthetic considerations are an important part of developing an appreciation for the beauty of the topic, both as an area of mathematics and through its applications to science, technology, and nature. This aesthetic appeal draws on both an appreciation of the power of the techniques and an awareness of the responsibility for a "tasteful" application of those techniques. The aesthetics approach is concerned with a selection of material that best develops an appreciation of mathematics.

Classroom Activities

We shall examine specific classroom activities that reflect the three distinct but interrelated aspects of the model and provide situations in which *statistical and probabilistic ideas are applied.* The activities are meant to highlight (1) the role of probability and statistics in everyday life; (2) the concepts needed to apply and interpret problems involving probability and statistics; and (3) an appreciation for the significant role of probability and statistics.

Most government agencies and business groups produce statistics. Different groups then attempt to interpret the data, often reaching different conclusions. For example, in examining unemployment statistics, some groups look at the data and conclude that the percentage is too small, since it fails to include people who are not actively seeking employment (they've just given up). Others will examine the same data and conclude that the percentage is too large, since it includes people who would be working for a second income to provide luxuries. The fact that such different interpretations do occur demonstrates the importance of teaching students to examine the assumptions underlying a set of statistical data *before* they interpret the results.

With the concern for statistical literacy among the population, the word *doublespeak* [51] was coined to describe the misleading use of data that is possible when many readers are statistically illiterate. (Numbers in brackets refer to items in the Bibliography.) Collecting a set of newspaper clippings is a valuable aid in the process of training "doublespeak detectors" and leads to many projects (see Projects). Students at all levels should be encouraged to collect and analyze their own data (articles 2, 3, 5, 7, 12, 16, 23). They could use data about their city, state, or province to predict future trends and, when the actual data become available, check the accuracy of their predictions. As they build a range for the prediction, particularly in situations where no obvious trend exists, they can begin to develop ideas of a confidence level and gain a sound base for later work (articles 25, 26).

At the high school level the same basic project could be undertaken with the major emphasis placed on the prediction and confidence aspects of the problem. Furthermore, the class could engage in a useful discussion regarding the basis for the statistical data: Who is included in the group? Do the data break down regionally, by town, and so on? What does *unemployed* mean? How does one gather data? The last question leads into a discussion of sampling procedures.

One set of data that is particularly valuable involves weather forecasting. These data not only allow for the collection and graphing of statistical information but can lead into a discussion of probability and the accuracy and significance of predictions (articles 11, 14). For example, consider weather information on rainfall. Most local radio and television stations give forecasts of the chance of rain for the next day. A discussion of what a 10 percent chance of rain means can be undertaken. A daily record can be kept of the predicted possibility together with the actual data on whether or not it rained. The students can then examine the accuracy of weather service predictions. Activities involving weather are particularly valuable for highlighting the idea of estimates made from a probabilistic base. It is impossible to be completely certain about weather forecasting, and thus we shall always be dependent on probabilities in this area. From these experiences, students follow naturally into such areas of discussion as how one assigns a probability to such disasters as nuclear radiation leaks, meltdowns, or the collision of oil tankers.

Advertising is another source for activities (article 23), particularly statements of the type that "four people out of five prefer product A." Find an advertisement of this type on a local radio or television station. Take the product to the class and divide the students into groups of five. For each group determine the percentage that uses the product. Discuss the results. Why were they different? Would it have been better to use a larger group? Is there a difference in the replies of male and female students? Then use the percentage given in the advertisement as a basis for comparison with the class. Such an activity provides experience with the importance of sample size and the nature of sampling (articles 25, 26).

The great suspicion with which polls are viewed during an election points to the need for a more effective way of teaching the ideas that underlie the process of polling opinions.

The concept of correlation frequently appears and is often associated with a search for causes and effects (see fig. 1.1). Is it fair to imply that the rise in crime and the rise in divorce rates are causally related? Cause and effect is another area that needs a more effective teaching strategy.

Government and business data lead to the development and practice of a variety of important aspects of probability and statistics. Census data contain a great deal of vital information. How accurate are these data? Some families fill out much more detailed forms. Why? Such questions lead to fruitful learning. New ideas, such as Lorenz curves showing income distribution, appear in government data and can be used to examine differences in the life-styles of various countries. Government data are of particular value in allowing emphasis to be placed on two key ideas, namely, (1) the *assumptions* underlying the gathering and interpretation of data and (2) the confidence that can be placed in these predictions. For example, consider another set of government data, the consumer price index (article

Costs and benefits

To mark the Queen's silver jubilee, the Economist, learned London weekly, has produced one of its admirable charts of social indicators—signs of how life has changed for the ordinary family in Britain.

Virtually all indicators of material life—from TV sets per household to emission of smoke per year—have improved during the last 25 years. To compensate for this, violent crimes have jumped 1,080 per cent and divorces 300 per cent.

We don't have indicators of similar quality in Canada, but we suspect that, if we did, the figures would come out close to Britain's and there's a sinister link between the increase in violence and the increase in divorce. That's the family. It's not disappearing but there's more tension pent up and hostility than we'd like to believe.

Our friends, Prof. Milton Friedman and his supporters, don't like the old concept of private wealth and public squalor, but if you redesign the concept in terms of prosperity and happiness, it makes us wonder about the true cost of economic development.

Fig. 1.1. An illustration of how statistics can be abused (reprinted from *Vancouver Province*, June 1977)

12). In calculating the index, researchers consider a variety of goods and services. These include housing, certain food items, and insurance. Thus, a four-point rise in the index affects different people in different ways. A four-point rise could result primarily from an increase in food prices, and in households where the food budget takes a substantial part of the week's expenditures, this increase would have far greater impact than in those where food costs are only a minor part of the expenditures.

If we are to educate students to handle the mass of statistical data being presented on television, radio, and in newspapers and magazines, it is essential that we teach skills that enable them to question and interpret data. The most recent trends in exploratory data analysis (article 16) point to the importance given by professional statisticians to the interpretation of data. These ideas, however, do not require advanced mathematical training. The development of these skills can begin very early in a child's school life. (See, for example, articles 5, 6, 9, and 16.) Many misconceptions can occur (articles 11, 14) if students are left to themselves to acquire an intuitive grasp of these concepts.

The second idea relates to confidence in predictions. It is vital that students realize the many important predictions such as those involving consumer prices (article 23). Inflation and weather forecasting are only probabilistic in nature; they are not certain. By building up intervals, students can start to obtain a feel for the accuracy of predictions. By examining predictions for monthly inflation rates, the students can come to realize that when an individual predicts an inflation rate of 1 percent for next month, nine times out of ten the actual increase will be between 0.7 percent and 1.3 percent (articles 25, 26). This intuitive feeling can later be formalized when students examine situations with specific theoretic distributions (article 21).

The area of sports and games provides highly motivating activities (articles 8, 10, 13). Most students go through phases in which they are interested in sports statistics such as hitting percentages in baseball, shots on goal in hockey, and yards rushing in football. These statistics provide an opportunity to examine the effect of individual "games" on concepts such as *average* (article 9).

The students could follow their favorite hitter's progress through a month. By starting at the beginning of the season, they would be able to see how one particularly strong or weak game early in the season can radically affect the hitter's average. As the season progresses, they see that individual games begin to have less effect on this average. Such activities help develop an appreciation of the affect of individual scores on means. This type of experience proves invaluable when a later, more formal discussion of concepts such as *mean, mode,* and *median* is undertaken and the effect of individual scores is probed. Furthermore, the predictive value of the  baseball player's average can be explored. What does an average of .333 say about the probability of a player making a hit in the next game or the next time at bat? Does the average provide a better prediction at the end of a month than at the beginning of the season? Is the average a good predictor of whether a player will make a hit in a game or in a series? Such questions build insights into the nature of prediction in probabilistic situations and reveal some of the paradoxes associated with averages (article 15).

Dice and card games are a useful source of problems for the discussion of probability. They also allow for the calculation of a theoretical probability for events (article 6). A sample dice game might divide the students into teams with one team scoring a point for an even throw and the other for an odd throw. The question is whether or not this is a fair game. Is it fair if one uses two dice and bases the scoring on whether the sum is odd or even (article 8)?

In throwing dice, one can find the distribution of 1s, 2s, ..., 6s for 30, 50, 90, 120, ..., 300 throws and see how the probability of each number changes as the number of throws increases. Which number of throws gives the "best" approximation to the theoretical probability? What would happen if the number of throws was increased? Would we obtain a better estimate? Such problems highlight the importance of sample size (articles 25, 26).

The greatly increasing complexity of today's world produces numerous problems having no exact solution. In such situations computers, simulation techniques, random digits, and statistics and probability all combine forces to produce approximations along with the probabilities of the levels of error (articles 16, 26, 28, 29, 30). To a mathematician these are valuable tools that give insight into many situations.

Most of the problems in this volume involve the student in such activities as analyzing assumptions underlying statistical data, exploring patterns hidden in such data, questioning the relevance and accuracy of inferences, and understanding the role of sample size and sampling methods. Teaching such key concepts is essential if we are to educate students in the use of statistical and probabilistic situations and give them a firm base for their later work. Activities that involve everyday life and draw on student interests help develop an appreciation for the role of statistics and probability.

II. Samples of Existing Courses or Programs

2. Experiential Statistics and Probability for Elementary Teachers

William A. Juraschek
Nancy S. Angle

ALTHOUGH mathematics educators have long advocated teaching statistics and probability in the elementary school, our observations suggest that these topics still are rarely covered at that level. We feel the best way to increase the likelihood of their being taught is to convince teachers of the accessibility and relevance of these subjects by presenting them in meaningful and novel ways. This was our objective in designing and teaching a summer course in statistics and probability for in-service elementary teachers.

The results of informal question-and-answer sessions (formal tests were omitted to minimize anxiety), together with unanimously enthusiastic responses on the course evaluations, convinced us that the course was a success. The consensus indicated that the teachers enjoyed the activities and planned to try them out in their own classrooms. Indeed, several have since reported successfully using many of the activities.

In preparation for the course, we surveyed the available materials, such as textbooks, curriculum projects, and the journal articles cited in the Bibliography. Recognizing that teachers usually teach as they are taught, we chose to use an experiential, informal, activity-based approach. For *probability* this meant many hands-on activities and simple experiments relating empirical and theoretical probability. For *statistics* we focused on descriptive statistics—especially graphs, tables, and charts—while including some intuitive ideas from inferential statistics.

Each of the four twenty-teacher sections met two hours a day for five weeks in an elementary school classroom. All materials, such as graph paper, marking pens, and dice, were provided. We used the first meeting to help the teachers get acquainted and to show them some basic graphing techniques. The first activity was called "Guess the Property." We had the

8

teachers separate into four groups of five. Each group was asked to brainstorm and select some property that could be used to differentiate the group members. Their property had to be something about themselves that was measurable and visible. When their property had been chosen, each group lined up in front of the class in order of increasing amount of the property and the rest of the class tried to determine the basis for the ordering. One group chose "width of smiles." Other choices included "length of sleeves," "amount of jewelry," "thickness of shoe sole," and "number of buttons." This entertaining bit of detective work was the springboard for an instructor-led discussion of quantitative and qualitative scales as well as discrete and continuous measurement.

Since we planned to have the teachers construct several graphs as part of their study of descriptive statistics, we used the remaining class time to introduce and model some graphing procedures. We chose grade level taught and number of years of teaching experience for data sets because the information was easily obtained and helped the class become better acquainted. It was pointed out that the two data sets had to be recorded in different ways. For the grade-level data we could logically form the seven categories, grades 1, 2, 3, 4, 5, 6, and other (some taught at several levels), before collection and then simply place tally marks in the appropriate category to obtain a frequency table. After a brief discussion of the types of graphs—bar, line, circle, and pictographs—we chose to construct a bar graph and a circle graph. We carefully constructed each graph on a large sheet of paper and displayed them for permanent reference.

In contrast to the grade-level data, the information about number of years of experience could not be categorized or grouped logically before collection; so we simply collected the data, recording all the numbers on the chalkboard. The range was large and the numbers diverse. After a brief discussion it became apparent that some grouping of the data would be necessary to produce a bar graph that conveyed the information effectively. A little more work and we had a model bar graph that displayed grouped data. At the close of the class we distributed the data collection sheets shown in figure 2.1 and asked each teacher to gather the requested information for use the following day. As you can see, the type of information requested is appropriate for elementary school teachers and pupils [see also 73; 98]. We made sure to avoid such topics as gross national products, world rice production, and the like, which do not usually spark great interest in the beginning statistics student.

At the second class meeting we had a large amount of data available. In each class, for example, we had the height and corresponding foot length of 60 adults, as well as the hair color of 100 persons. This provided several good-sized data sets to graph and analyze. We did not feel it was necessary for each teacher to graph each data set, and so we had them work in pairs to collect the class totals for three data sets and then display the informa-

Data Collection Sheet

As directed by your instructor, collect and record the following data. Your results will be combined with those of your partner and the rest of the class. With the help of your instructor, you and your partner will display selected data sets using tables and graphs. Unless told otherwise, don't use your classmates in your sample.

1. For three adults, record their height and length of bare right foot to the nearest centimeter.

	#1	#2	#3
Height	___	___	___
Foot length	___	___	___

2. Record the favorite ice cream flavor of five persons including you. There are two ways; try both.

   First, simply ask for their favorite flavor.

#1	#2	#3	#4	#5

   Second, ask for their favorite among chocolate, strawberry, vanilla, and other.

#1	#2	#3	#4	#5

3. Record the eye color of five persons including you.

blue	brown	gray	other

4. Record the hair color of five persons including you.

blond	brown	black	red	gray	other

5. Record the number of letters and spaces in the names of five persons including you. (The name used to sign checks.)

6. Record the number of each type of coin in your possession right now.

1¢	5¢	10¢	25¢	50¢	$1

7. Record the favorite TV program of five persons including you.

8. Record the birth month of five persons including you.

Fig. 2.1

tion using tables and graphs. To ensure variety, no two pairs were responsible for the same three sets (fig. 2.2).

Fig. 2.2

As the teachers collected their data and planned their graphs, we circulated among them answering questions and offering suggestions. The finished products, many of them pleasantly creative and eye-catching, were displayed around the room for celebration and future reference (fig. 2.3). The variety of styles and types of graphs excited further discussion of which displays best fit which types of data. This was especially so for the scatterplot of the data on height and foot length in data set #1 (fig. 2.1). Many of the teachers had heard of correlation but were surprised to "see what it looks like."

One important discovery from the graphing exercises was that whereas some graphs are more effective than others, there is no one best way to graph information. Sometimes a bar graph is just as effective as a circle graph, as in data set #6. Another observation was that some data sets do not lend themselves easily to graphical representation. For example, in data set #7 very few TV programs were the favorite of more than two persons, and although this great variety was visible on the graph, the scales of the graphs were so cumbersome as to be confusing. The teachers also found that the way a question was asked made a difference in the data obtained. As they collected data for set #2, they realized the difference between an open-ended and a multiple-choice format. It was becoming apparent to these teachers that statistics might not be such a cut-and-dried subject after all.

Fig. 2.3

While they were collecting, organizing, and displaying their data sets, the teachers were also reading assigned materials about constructing graphs and tables [88; 98]. Thus, by the end of the first week they had gained both academic and experiential knowledge about processing data. It was now time to try something more experimental, something we hoped would lead to some elementary ideas about making inferences from sample data.

Since both adults and children are familiar with automobiles and automobile traffic, we asked the teachers to conduct a traffic survey designed to answer some questions related to autos and traffic. The only constraint was that the data be collected on a city street during a fifteen-minute period. To facilitate brainstorming and data collection, the teachers worked in groups of four, each group designing a survey to answer its own question or questions: What is the average number of persons in each auto? What is the frequency and distribution of car colors? What proportion of cars are using air conditioning? What is the frequency distribution of regular and compact cars? How many cars run yellow lights?

The teachers quickly discovered that many decisions were necessary, both in collecting and in interpreting the relevant data. For those tabulating car color, for instance, problems of categorizing rapidly emerged. Should an aqua car be counted as blue or green? How many color categories should be used? How should a two-toned car be classified? The group counting car occupants found a mean of 2.2 persons per auto; as they conjectured on why this differed from the city average of 1.1 cited in government surveys, they became aware of the idea of a representative sample and the importance of choosing one. (In this example the sample was taken on a busy neighborhood street at about 10:00 A.M., and the cars seemed to contain a disproportionate number of mothers with children.)

The group that counted the number of cars using air conditioning initially planned to display a sign that read "Are you using your air conditioner?" and record responses. However, after some discussion they perceptively conjectured that any car with its windows rolled up on such a hot summer day must be air conditioned and if so, this tendency would allow for less obtrusive and easier data collection. To test their conjecture, they polled cars stopped for a red light and found that nineteen of the twenty with windows rolled up were using their air conditioners. (The light changed before they could determine why the one driver without air conditioning had the windows up!) From this little survey they concluded that their conjecture was sound.

On completion of the surveys, each group displayed their information and discussed their experience. It was readily apparent that not only had the teachers improved their graphing skills but they were becoming involved in genuine problem solving, creatively analyzing data, and testing the validity of generalizations. Moreover, they were enjoying the process!

Although we as instructors chose the course content and most of the activities, our role was more that of resource consultants than of traditional professors. From the outset we stressed processes, giving a minimum of direction and providing answers only when we felt they could not be discovered eventually by the teachers. This approach, far from typical of that observed in most elementary classrooms, generated some anxiety at first, but the teachers soon developed generally independent problem-solving attitudes as well as confidence in their own abilities. The few who remained timid followers were the exceptions. Learning and discovering had become an adventure, sometimes difficult but always rewarding. During the second week, when class time was devoted primarily to lectures and discussions of frequency distributions, central tendency, and variation, thereby laying a foundation to link probability with statistics, the teachers were actively involved. As we referred to the data sets and graphs from their work of the previous week, material with which they were very familiar, their questions, conjectures, and observations revealed both increasing interest and understanding.

As part of our focus on process, we emphasized the interplay between experiences in doing mathematics and the theory of mathematics. We introduced this interplay with a series of experiments in tossing coins [see also 115; 132]. These experiments are described here in some detail because they are so representative of the general method and philosophy of the entire course.

Each teacher flipped a coin ten times and recorded the number of heads. Class totals showed that very close to half the 200 tosses had been heads, although many of the samples did not have exactly five heads. We discussed the difference between expectation, or expected frequency, and what actually occurred. Most people had no difficulty accepting that although the expectation is for heads 50 percent of the time, a small sample often does not come out that way. This activity served as an intuitive example of the law of large numbers.

Next, the class worked in pairs, and each pair flipped two coins simultaneously, recording the results as "two heads," "two tails," or "mixed" (one head and one tail) for twenty trials. It became clear as the class results were compiled that many of the students expected an approximately even distribution among the three categories and were surprised when the mixed category had a much greater frequency than the other two. Since the empirical results came first, the teachers had no theoretical comparison for their expectations. Some of them greeted the explanation, presented by means of a tree diagram (fig. 2.4), as a welcome relief for their confusion. The fact that the mixed category actually contains two separate outcomes—head-tail and tail-head—is not obvious but emerges from both the data and the diagram. The theoretical reason for the outcome was presented at a level the teachers easily understood, following the collection of empirical data.

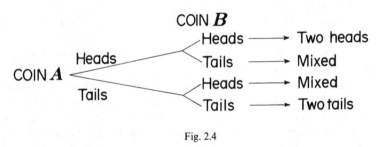

Fig. 2.4

In almost all instances this was our approach. If we wanted to present the theory of some underlying abstract concept to the class, we did an experiment first so that they had an experiential basis for the abstraction. The mathematical rules governing the assigning of a probability to an event seem to be much more comprehensible when this order is followed.

We then extended the experiment to three coins. By this time the teachers had guessed that the outcomes for the four categories—three heads, two heads and one tail, two tails and one head, and three tails—would not be equal. Some were able to find the expected frequencies, using a tree diagram, before the data from the experiment were collected. Again, each pair recorded the results from twenty trials, and class totals were displayed on the chalkboard. Very quickly, one of the teachers would make the observation that adding a coin doubled the number of possible outcomes so that now there were actually eight possibilities, although we had grouped them into four categories.

We had planned to end the coin-tossing activity at this point, but the teachers were interested and involved. In every section they asked what would happen if four or more coins were used. They were told either to try the experiment or construct a tree diagram before the next meeting. By the beginning of the next class, many of the teachers had worked with the problem and were eager to share their results. By using a tree diagram to list the sixteen possible outcomes with four coins, the class was able to assign probabilities to all categories of outcomes. Several of them remarked that as the number of coins and possible outcomes increased, both the diagrams and the management of the empirical method became more cumbersome. What would one do if one wanted the probabilities for many coins? One teacher recognized the numbers that had turned up and, although she could not name it, described Pascal's triangle. Figure 2.5 shows the final result of a table that evolved on the chalkboard through the combined efforts of class and instructors.

Number of coins	Number of possible outcomes	Number of outcomes in order from all heads to no heads						
1	$2 = 2^1$				1	1		
2	$4 = 2^2$			1	2	1		
3	$8 = 2^3$		1	3	3	1		
4	$16 = 2^4$	1	4	6	4	1		
5	$32 = 2^5$	1	5	10	10	5	1	

Fig. 2.5

The class saw the pattern of entries and could generate new rows easily. They also found that they could use the table to establish theoretical probabilities. For instance, the probability of five heads in five coins is 1/32, whereas the probability of three heads in five coins is 10/32. Many of these teachers, who had previously learned about Pascal's triangle in an algebra class, were delighted to find a way in which it really could be used, and in a form that made sense to them. As they developed the table, most of the class became involved in a discussion of the binomial distribution, and this led to a consideration of the normal distribution. Originally we had omitted this material as being too theoretical for their needs.

This series of experiments with coins set the pace for the remainder of the course. We did a similar sequence of activities with dice and even taught the teachers how to shoot craps—an obvious treat for most of them. With samples drawn from decks of playing cards, we introduced concepts such as sampling with and without replacement, independence of events, and mutually exclusive events. The teachers determined and compared both theoretical and empirical probabilities in several activities involving cards, coins, and dice.

At this point we conducted a series of activities for which it is impossible, or at least difficult, to assign theoretical probability. In one of these, each teacher was asked to toss ten thumbtacks and record how many landed point up [see 35; 115; 132]. Before any tacks were tossed, the class speculated about the relative frequency of point-up tacks and soon discovered that such an a priori assignment of probability was clearly impossible. The only sensible way to obtain an estimate of this probability, aside from a complex analysis of the vector physics of a thumbtack, was to toss the tack. This we did, constructing a cumulative relative frequency graph like that in figure 2.6. Each teacher added his or her point-up frequency to the total, and eventually the cumulative relative frequency approached a fixed value. The teachers agreed that this would be an acceptable estimate of the probability that a tossed thumbtack lands point up.

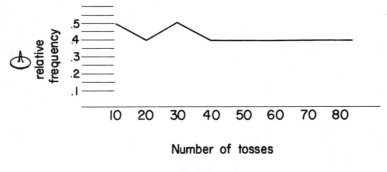

Fig. 2.6

For a second experiment like the one above, and as an extension of those involving coin tossing, we asked each teacher to bring three sturdy cylinders to class. Most brought clean metal cans with ends and labels removed, but some brought heavy cardboard tubes, oatmeal boxes, or paper towel tubes. The cylinders were marked "heads" at one end and "tails" at the other. The teachers classified each cylinder with a number that was the ratio of height to diameter; a tunafish can might be labeled 0.3, and an oatmeal box might be labeled 1.8. Next the teachers flipped the cylinders twenty times and recorded the frequencies of "heads," "tails," and "side" [see also 132]. As with the coins, the cylinders came to rest on heads about as often as on tails, but a clear trend emerged with respect to the side frequency. The cylinders with larger ratios landed less frequently on end and more often on their sides. The tuna can, with a ratio of 0.3, seldom landed on its side, while a cardboard tube with a ratio of 3 had only a very small probability of landing on either end. One teacher quickly measured a nickel, found a ratio of 0.05, and suggested that that explained why the coin "never" lands on edge when flipped.

We chose to describe our last two activities because they simulate real-world applications of sampling techniques. In the first, each teacher randomly selected a page from a book, magazine, or newspaper; chose a passage of about a hundred words, and tallied the frequency of each letter of the alphabet. Since the data were rather cumbersome, we collected class totals by having pairs of teachers add their results, then two pairs combine their totals, and so on until class totals were found. These were written on the chalkboard and relative frequencies computed. The class discussed how the distribution of the letters in the English language might affect printers, signboards, and telegraph messages sent in Morse code [see 78]. We suggested that elementary teachers and students might want to conduct a study to determine whether or not these relative frequencies remain constant in reading material across grade levels.

A second simulation activity was based on the sampling techniques used by pollsters to gather data about large populations [115]. We gave each pair of teachers a paper bag filled with a number of colored cubes, instructing them to reach in blindly, extract a sample of predetermined size, record the colors in the sample, and return the cubes to the bag. After the teachers shook the bags to distribute the cubes randomly, they repeated the procedure three more times. From those samplings they estimated what color cubes, in what proportions, were in the bag. Each pair completed this experiment for five different bags, listing their estimates for comparison by the class. Class discussion led to a consensus estimate for each bag, which was promptly tested by dumping out and counting the contents. Almost every estimate was close, and many were exact. This activity generated a heated discussion of TV rating systems and polls that affect our lives.

An activity that fell short of expectation was, nevertheless, a source of much interest and entertainment. Intrigued by stories in the popular press, we hoped to motivate the class to study some elementary inferential statistics with an experiment on the effect that talking to plants has on their growth rate. At the second meeting we asked the teachers, in pairs, to plant two nasturtium seeds in paper cups. Over the next four weeks, they were to care for their seeds in the classroom under identical conditions, except that each pair would talk daily to the plant in the experimental group and not talk to the plant in the control group. Excitement spread when the first few seeds sprouted. Soon everyone had some sprouting plants to care for as well as growth curves to plot. All went well until the teacher-botanists returned from their four-day Fourth of July weekend to find, sadly, that the intense heat in the closed classrooms had taken its toll. Although our ultimate goal was not attained, the teachers did enjoy comparing the growth curves and speculating about how things might have come out.

We offer the following guidelines for choosing and designing learning activities for in-service teachers. Each activity should (1) be relevant and interesting to classroom teachers; (2) be of appropriate mathematical sophistication; (3) involve readily obtainable data; (4) encourage various solution strategies; and (5) be easily modifiable for use with children.

Our experience has supported the assertion that such activities nurture mathematical confidence and stimulate learning among elementary teachers and, moreover, are likely to find their way into elementary school classrooms.

3. Teaching Statistics to Eleven-to-Sixteen-Year-Olds

An Account of Some Development Work in England and Wales

Peter Holmes
Daphne Turner

WHAT IS *statistics?* Any standard introductory text gives two basic definitions with which teachers of statistics will be familiar: Statistics is (1) a collection of numerical data and (2) the science of handling that data. But the same question asked of most school students who have studied statistics will probably produce neither of these answers. They are more likely to refer to the techniques involved—drawing a bar chart, finding a mean, calculating a standard deviation. Yet no scientist would regard a list of techniques as reflecting fairly the teaching of science. Why, then, should a teacher of statistics?

All too often, however, statistics *is* taught as a series of techniques. In mathematics lessons the subject is often seen as little more than an extension of arithmetic, with algebraic substitution included for good measure. In biology, for example, students need to use a standard deviation, and so the appropriate formula is given and numbers substituted. The question, "Why the standard deviation?" is seldom discussed.

Applications of statistics in everyday life are not difficult to find. Indeed, the average citizen in any developed country, and increasingly in the developing countries, needs to be aware of the why, how, and where of both data collection and processing. An understanding of why statistical data are collected can, in the end, lead to much more accurate data through the sympathetic cooperation of those whose job it is to provide it. A knowledge of where and how data are used leads immediately to a greater understanding of the need for a census, the purpose of retail price indexes, and the use of such summary statistics as the average family.

Seldom, however, is the student made aware of the background *thinking* behind such figures. Given well-rounded, easy data, they can, no doubt, compute a simple mean or a weighted average to obtain a hypothetical price index, but what of the real-life situation? Where do the *real* figures come from? Given a mass of data, how does anyone set about tidying it up into something manageable and understandable? These questions, surely, are what teaching statistics should aim to answer.

The Schools Council Project and Its Approach

The consideration of these questions has affected recent developments in statistical education in England and Wales. In the early seventies deficiencies in the teaching of statistics became apparent to many professional statisticians; teachers also expressed their concern but felt themselves in need of more help and guidance to overcome the problems. Thus a curriculum development project on statistical education was established and funded by the Schools Council for Curriculum and Examinations. Included in the aims of the project was devising detailed proposals for the implementation of teaching statistics to all students between the ages of eleven and sixteen as part of their general education. (The original three-year project has since been extended.)

From the outset it was obvious that a totally fresh approach to teaching statistics was needed. To achieve this, a problem-solving approach was adopted throughout the material produced by the project team. It was strongly felt that students would profit from seeing the *need* for statistical techniques—why, say, a mean or a pictorial representation was used at all. How the mean was obtained or the best way to represent the data was no less important, but the *how* was to be clearly seen as a development from the *why*.

One hurdle faced by the team was the need to establish what students in the eleven-to-sixteen age range should be capable of learning in the field of statistics. Thus an early part of the project's work was to produce a hierarchies paper that not only established an order of priorities but also aimed to give an overall perspective to the development of statistical concepts and techniques throughout this age group. This paper formed the basis for the teaching materials, which aimed to develop the concepts and techniques in a practical context. The inability of students to understand fully the theoretical justification of some particular concept did not necessarily mean that it was excluded—only that the basic idea would be covered by experiment so that students would achieve at least an intuitive idea of, say, a likely outcome.

The teaching materials

Altogether twenty-seven booklets (or units), each at one of four levels, have been produced for pupils' use, each one accompanied by teacher's notes. Not all topics are applicable solely to mathematics teaching: some are included for such disciplines as the humanities, science, and social science. Yet in all the units a problem is posed for pupils, data are obtained to help solve the problem, and the introduction of statistical techniques and elementary inferential statistics follows naturally.

Levels 1 and 2. Units in the first two levels have been designed to lay the foundations for understanding probability and the use of statistical techniques. One first-level unit asks students whether a 6 is harder to get than any other number when throwing a die in a game of chance. Some experiments then enable students to test their conjecture systematically. The data thus obtained are rather difficult to comprehend, since many figures are involved. This suggests that pictorial representation and data reduction may be necessary, and so students are encouraged to draw bar charts and calculate averages—the median and the mode.

Once it has been established that with a normal die the probability of obtaining any one number is the same as for any other, the students are asked in another unit, "How can you tell whether a die is biased or fair?" A simple dice game is then introduced to allow them to investigate this problem. Basic ideas of probability in other contexts are introduced in similar ways. By the end of this unit students have compared relatively complicated probabilities without carrying out any detailed calculations. Hence an intuitive feeling for probability is developed.

Second-level units continue to reinforce the learning of statistical techniques through the use of problems. One is concerned with the writing styles of different authors. The initial problem posed is, "How can we recognize style?" This is followed by the question, "Can you identify the author from his writing?" To answer these questions, the students use pictorial representation and measures of central tendency and spread.

Questionnaires are used to collect data for various purposes, both in and out of school. Many of the limitations of questionnaires, however, are not evident to the writer of the questions; it is only when the questions have to be answered or the responses analyzed that the weaknesses become apparent. This is particularly true of questionnaires designed by students in many curriculum areas of their school program. One unit produced by the project team, called "Opinion Matters" and tested with twelve-year-olds, looks closely at the design of questionnaires and the effect of their wording on the usefulness of the replies. The unit starts by asking students to answer some badly worded questions, including such vaguely worded ones as *How many sweets do you eat?* and *Mother's age at birth?* They are then given the task of rewording the questions so they can be answered. Having seen some of the difficulties, the students are more cautious when rephrasing and are better able to assess critically their own and other students' attempts.

The effects of wording on the responses obtained and also on the ease or difficulty of replying can be shown by using two similar questionnaires, X and Y. Ostensibly, both are designed to elicit the same information, but as figure 3.1 shows, there are subtle differences. The class is divided into two teams, each member of team 1 receiving a copy of X and each member of team 2 receiving a copy of Y. Each team sees only one questionnaire. The forms are completed and class results are obtained.

To test the effectiveness of the wording, trial schools were invited to return class responses along with their evaluation forms to the project team. The results of some of the questions from one school can be seen in figure 3.1.

	Yes	No	
X: I'd rather be a humble doctor than a successful pop singer.	8	22	
Y: I'd rather be a doctor than have to sing for my living.	20	10	
	Yes	No	Don't know
X: Should all pupils be allowed to stay at school until they are 16?	14	8	8
Y: Should all pupils be forced to stay at school till 16, even when they don't want to?	8	20	2

Fig. 3.1. Effect of wording on responses

Another question, common to both questionnaires, illustrates the futility of asking questions beyond the understanding of those responding. All thirty-two responses to the question, "Should differential calculus be taught logically rather than genetically?" were "Don't know."

Reaction from teachers to the unit was mixed, their main reservations being (1) the time required to study the unit completely and (2) the question of where such a topic could best be placed within the school curriculum. Nevertheless, it made teachers as well as pupils more aware of the problems involved. Among teachers' comments was, "It provided good discussion, bringing out ideas like *emotive language* and *bias.*" Overall there was general agreement that such a unit should find a useful purpose within a school; one teacher commented: "The final analysis will be after the pupils have left school and are part of the outside world. . . . An essential part of life is to be critical."

In completing this unit, students learn to spot errors of design and encounter general principles to be followed in writing good questions. This helps them appreciate more fully the limitations of opinion polls reported in newspapers, an essential part of basic numeracy (i.e., numerical literacy).

In the process of working through the units students gain experience in handling both their own and other people's data. They learn to appreciate not only how small samples sometimes exhibit abnormalities but also how, by pooling their results with those of other class members, increasing the number of trials can lead to a relative frequency that matches more closely a theoretical probability.

Levels 3 and 4. After completing the second-level units, students will have met most of the elementary statistical techniques; it is then possible to apply them in a wider context while still extending the students' statistical knowledge. From the outset the project team felt it essential that the material produced should be seen to be relevant to everyday life and that it should help meet some of the needs of teachers using statistics in subject areas other than mathematics. These considerations were borne in mind when producing the materials, particularly at the third and fourth levels.

One third-level unit, relevant to geography, economics, and social science, uses the themes of populations to investigate simple projections and the possible inaccuracies that may result from their use. Subsequent sections illustrate such demographic techniques as population pyramids and the effects of birth and death rates on population growth.

Another unit, relevant to many subject areas, considers the abuses of statistics. When students have studied the unit, they should be better able to criticize constructively any of the more common misuses of statistical data in arguments or advertisements.

Simulations are used frequently throughout the units. One third-level unit uses a simulation to investigate how scientists can estimate the number of whales in a particular ocean; a fourth-level unit uses one to study the effect of chance both in a taste-testing competition and in multiple-choice examination questions. Thus students are made aware of the implications of statistical theory to their own immediate situation.

Other fourth-level units are particularly applicable in the fields of economics and related subject areas. The unit "Figuring the Future" investigates the changing pattern over a number of years in the use of television, letter post, and telecommunications, again considering the possible limitations of the use of trend lines to predict the future. Another unit considers how a retail price index attempts to reflect the spending habits and the changes in the cost of living of an ordinary citizen.

One unit that has proved particularly popular at this level and of use and interest to older students as well deals with the apparent effects of maternal smoking on birth weight and the possible relationships between deaths from particular illnesses and the number of cigarettes smoked. In these ways students come to appreciate the applications of statistics in the real world.

The design of the units allows students to work at their own pace with occasional breaks for the pooling of ideas, reactions, and data—and for discussion. Optional sections are included in all units to serve one of two purposes: either to give additional practice if needed or to provide additional stimulation for more able students. The use of calculators is not essential to the work but is at times advised to facilitate working with cumbersome real data.

Reaction to the Material

Evaluation of the material was considered very important. All the units produced by the team were tried out in a variety of schools in England with classes of varying ability levels. This enabled comments from teachers and students to be considered, and in light of the reactions appropriate amendments were made to both the format and the content of the units. Sometimes the rewriting of the final versions was carried out by the teachers themselves, preserving, of course, the problem-solving approach. The outcome has been a set of well-tried and well-tested units.

The material has been received warmly by many, particularly students. Although some found the idea of discussion and open-ended questions very different from much that they were used to in mathematics lessons, they were quick to adapt, and many enjoyed their lessons all the more because of the different approach. One interesting feature was that those who often had great difficulty in the normal mathematics lessons were able to make a useful contribution to any discussion on the implications of the data with which they were faced.

Even though the team hopes the materials will be readily accepted in published form, the production of teaching material does not necessarily mean that all teachers will be willing to change the habits of a lifetime. During the trials it was obvious that although teachers were prepared to test the material, not all of them were committed to adopting it in the way the project team felt was most appropriate. Many found themselves too constrained by the limitations of the external examination system, which the project has only just started to influence to any great extent. Others felt that such an approach was too time-consuming, and that the traditional approach achieved more in the time available. What they failed to realize was that the end results were not expected to be identical. Whether the traditional or the problem-solving approach was adopted, the students should, at the end, have some mastery of statistical techniques; however, with the problem-solving approach, all students, whether specializing in statistics or not, should have an understanding of "Why statistics?"

The time remaining to the project team has been assigned to disseminating the materials in an attempt to widen the teaching profession's acceptance of the need to approach statistics in the problem-solving way and to make teachers from one discipline more aware of the statistics being taught in other disciplines.

Do *you* know the extent of the statistics taught in the science, geography, or economics departments of your school? If you don't, may we suggest that you make some inquiries? You could be in for quite a shock. It may be that if all the time your students devote to statistics during their school life were considered as a whole, there would be plenty of time for a problem-solving approach. We are convinced that all students would benefit.

4. An Elementary Course in Nonparametric Statistics

Albert M. Liebetrau

Courses that complement traditional mathematics courses are being offered by secondary schools in increasing numbers. Among these, a well-planned course in probability and statistics is one of the most important for several reasons. A course in statistics enables students to improve their mathematical skills while simultaneously exposing them to many practical problems to which these skills can be applied. By its very

---

In addition to references cited in this article, [23], [83], and [120] are particularly sympathetic to the attitude expressed here. All are excellent source books for anyone teaching an elementary course in probability and statistics.

nature, a study of statistics provides an excellent opportunity for students to pursue individual interests that cut across traditional subject-matter boundaries. Any statistical decision problem is like a gem with multiple facets: the mathematically inclined student may find it interesting to derive the various probability distributions and their properties from theoretical considerations; computational questions exist to challenge those interested in computing; and, of course, the problem itself can arise from almost any source.

The course proposed here is for students with at least two years of high school algebra. The presentation of most topics, however, can easily be adjusted to accommodate students with different levels of mathematical skill. Most of the ideas can be mastered by students with less than two years of mathematics, and it is possible to present topics with sufficient depth to challenge those with more advanced mathematical training.

The course differs in two significant ways from the usual courses in statistics offered at this level. The first concerns the treatment of ideas from probability. Probabilistic foundations of statistics are extremely important and *should not be omitted.* It is essential for students to realize that various statistical methods depend on certain assumed probability models. Otherwise, students are left with a vague and imprecise concept of probability that provides no insight into how statistical computations are actually carried out in real problems. Although ideas from probability should not be ignored, neither should excessive emphasis be placed on counting problems, since the truly important ideas involve the properties of probability models and random variables. Combinatorics should be considered only insofar as necessary to develop models for sampling with and without replacement. The treatment in chapters 3–5 of [4] focuses on these ideas without overemphasizing combinatoric aspects.

A second important feature of the proposed course is the emphasis on graphical and nonparametric methods. Classical statistics depends heavily on the normal (Gaussian) distribution, whereas nonparametric methods are developed under less restrictive assumptions about the distribution of the data. Classical methods depend on properties of the normal distribution, and these can be quite complicated. Nonparametric procedures, however, are usually based on rankings or permutations of the data. As a result, many nonparametric methods, in addition to being mathematically simpler than their classical counterparts, have greater intuitive appeal also. Furthermore, nonparametric tests are usually more powerful in situations where the data do not follow the normal distribution. The fact that non-normal data occur so often in practice is a strong argument for an increased use of nonparametric methods.

An outline of the course appears in figure 4.1. Following the outline are two sections in which selected topics not usually found in elementary statistics textbooks are presented in a manner suitable for classroom use.

Outline of a Course in
Nonparametric Statistics

I. Topics in probability
   A. Sample spaces and probability measures
   B. Random variables and some of their properties
      1. The probability function
      2. Expectation (mean) and variance
      3. The cumulative distribution function
   C. Two or more random variables
      1. The joint probability function
      2. Covariance and correlation
      3. The mean and variance of sums of random variables
   D. Important probability models
      1. Models for counts
         a) The binomial distribution (sampling with replacement)
         b) The hypergeometric distribution (sampling without replacement)
         c) The geometric distribution (waiting times)
      2. Continuous models
         a) The uniform distribution
         b) The exponential distribution
         c) The normal distribution
   E. The *central limit theorem:* using the normal distribution to approximate other distributions
II. Topics in statistics
   A. Examples of statistical decision problems
   B. Empirical data analysis: informal and graphical techniques
      1. Histograms and stem-and-leaf plots
      2. Quantile-quantile *(Q-Q)* plots and probability plots
      3. The sample cumulative (empirical) distribution function
      4. Scatter *(X-Y)* diagrams
   C. Toward more formal statistical procedures
      1. Populations and samples
         a) Random, representative, and stratified sampling
         b) Relationship between a sample and the underlying probability model
      2. Some important statistics
         a) Sample moments
         b) Some (very) elementary properties of order statistics
         c) Sample quantiles and the empirical distribution function (again)
   D. Quantitative data analysis: methods based on statistical models
      1. A first look at testing: the binomial case
         a) Hypotheses, significance levels, and power
         b) The sign test
      2. Inferences for one sample
         a) Median and quantile tests
         b) The signed-rank test
         c) A test of randomness versus trend
         d) Goodness-of-fit: the chi-square and Kolmogorov-Smirnov tests

3. Inferences for two samples
   a) The sign test (again)
   b) Runs tests
   c) Procedures based on ranks
   d) Permutation tests
   e) The chi-square test of independence
   f) Correlation: the rank-correlation and product-moment correlation coefficients

Fig. 4.1. Outline of the course

Graphical Techniques: Stem-and-Leaf Plots and Q-Q Plots

A great deal can usually be learned about a set of numbers simply by plotting them in different ways. One should always look at the plots of data one studies because plots often reveal relationships that are hard to see in any other way. A stem-and-leaf plot is similar to a histogram but contains more information about the data. The Q-Q (quantile-quantile) plot is useful for comparing two populations or samples.

To illustrate the use of these plots, suppose we look at the age at death of former U.S. presidents. We shall form two groups, the first consisting of George Washington to Andrew Johnson and the second consisting of Rutherford Hayes to Lyndon Johnson. We shall omit Ulysses Grant, since the Q-Q plot is easier to make if the two groups are of equal size. Thus, we have two groups of data, each consisting of the ages at death of former U.S. presidents:

First group: 67,90,83,85,73, 80,78,79,68,71, 53,65,74,64,77, 56,66
Second group: 70,49,57,71,67, 58,60,72,67,57, 60,90,63,88,78, 46,64

Consider the second group. To make a histogram, we would split the range of data into several categories and plot an X in the appropriate category for each observation. It is just as easy, and more informative, to let the first digit (stem) determine the category and write the second digit (leaf) adjacent to the first but separated from it by a vertical line. Thus, 70 has 7 as a stem and 0 as a leaf. The stems (listed here in order of occurrence)—7, 4, 5, 6, 9, 8—are shown in figure 4.2(a). The stem 7 has leaves 0 (from 70), 1 (from 71), 2 (from 72), and 8 (from 78); these are shown in (b). We continue to write the leaves associated with other stems until all have been plotted, as in figure 4.2(c). Finally, the finished stem-and-leaf plot in (d) results from ordering the leaves of each stem.

A stem-and-leaf plot has several nice properties. First, we do not have to make arbitrary categories as we must for a histogram, because we let the data do it for us. Second, we can recover all the original data from a

```
9|         9|          9|0         9|0
8|         8|          8|8         8|8
7|         7|0128      7|0128      7|0128
6|         6|          6|707034    6|003477
5|         5|          5|787       5|778
4|         4|          4|96        4|69
(a)        (b)          (c)         (d)
```

Fig. 4.2

stem-and-leaf plot because we have written each digit (leaf). This is impossible from a histogram, since all observations have been replaced by a single symbol. Furthermore, a stem-and-leaf plot gives the same visual impression as a histogram and certainly can be constructed with no more difficulty. Finally, we shall see later that a stem-and-leaf plot is quite useful for ordering data.

There are many variations of the simple stem-and-leaf plot we have just completed. We can, for example, split each stem into two parts (shall we call them twigs?) by writing the leaves 0–4 on one line and 5–9 on another. Thus where "." and "*" are placeholders for the digits 0–4 and 5–9, respectively. Many other possibilities are discussed in [136]. Computer programs to do the work can be found in [80].

Now let's turn to the question of comparing the first group with the second. We can do this with side-by-side plots like those in figure 4.3(a), but it is usually better to make back-to-back stem-and-leaf plots as shown in figure 4.3(b). Again, there are many variations.

```
9|0          9|0               0|9|0
8|035        8|8             530|8|8
7|134789     7|0128       987431|7|0128
6|45678      6|003477      87654|6|003477
5|36         5|778            63|5|778
4|           4|69               |4|69

  First        Second          First   Second
      (a)                          (b)

 Side-by-Side                 Back-to-Back
```

Fig. 4.3

Are you surprised to see that our earlier presidents lived considerably longer than those who took office more recently? What plausible explanations can you give?

Although a back-to-back stem-and-leaf plot is helpful for comparing the *locations* of two distributions, a two-dimensional plot gives better resolution for comparing their *shapes*. Suppose the observations in each group are ordered:

First group: 53,56,64,65,66, 67,68,71,73,74, 77,78,79,80,83, 85,90
Second group: 46,49,57,57,58, 60,60,63,64,67, 67,70,71,72,78, 88,90

(Notice how easy it is to order the data with the aid of the stem-and-leaf plots in figure 4.3!) We now plot the pairs of observations with corresponding ranks on axes with a common scale to get the *Q-Q plot* shown in figure 4.4.

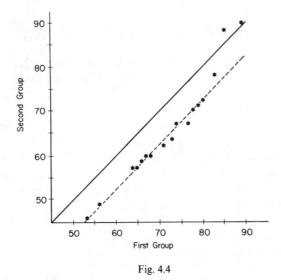

Fig. 4.4

If the two groups of data come from the same distribution, we expect most of the points to fall quite close to the 45° line. Instead, we see that the points (except for the last two) form a straight line about eight units to the right. This tells us that the two groups have distributions with (almost) the same shape but with one offset from the other by eight units. Actually, the sharp rise of the last two or three points from the dotted line shows that the upper tail of the second group is longer than the upper tail of the first. Notice how much easier this is to see from the *Q-Q* plot in figure 4.4 than from the stem-and-leaf plot in figure 4.3.

The plot in figure 4.4 is called a *Q-Q* plot because quantiles from one data set are plotted against corresponding quantiles from another. The reason for starting with data sets of equal size was to make the calculations as easy as possible, since then we need only pair corresponding (ordered) values. It is a bit harder to compute corresponding quantiles when the sample sizes differ, but once this is done, the *Q-Q* plot is constructed exactly the same way as the one in figure 4.4.

Like stem-and-leaf plots, Q-Q plots can be modified in various ways. For example, data values could be plotted against the corresponding values we would expect, *provided* the data *actually* come from a particular distribution we have in mind. A straight line would confirm our choice, whereas some other shape would give evidence to the contrary. If we have the normal distribution in mind, the resulting graph is called a *normal probability plot.*

Good references on graphical methods are widely scattered in statistical literature. Although some graphical techniques are discussed in [4] and [80], [136] is more comprehensive.

Nonparametric Rank Tests: The Rank Sum Test

Statisticians find in practice that certain basic questions arise repeatedly. We shall look at one such question and discuss in some detail a rank test for dealing with it.

From a statistician's point of view, all of the following are two-sample problems because all require us to decide whether or not two populations, say X and Y, have the same distribution. Do two different drugs have the same effect on tumor growth? Do boys and girls of a given age perform equally well in mathematics? Do cars from the current model year get better gas mileage than comparable cars from the previous year? We shall use the last example to show how the rank sum test can be used to help us decide whether or not new cars get improved gas mileage. (We shall assume that all distribution functions are continuous so we do not have to worry about calculating the probabilities of tied ranks.)

Consider the claim of an automobile manufacturer that the new models get better gas mileage than corresponding models from last year. Suppose $m = 3$ cars from last year and $n = 4$ from this year have been carefully measured to determine gas mileage and that these data result:

Previous year (X): $X_1 = 17.4$, $X_2 = 21.0$, $X_3 = 18.6$
Current year (Y): $Y_1 = 20.2$, $Y_2 = 23.6$, $Y_3 = 24.3$, $Y_4 = 21.9$

Here, X_i represents the ith observation on X, so that X_2 is the second X value, Y_3 is the third Y value, and so on. Unrealistically small sample sizes have been chosen so that calculations can be given in detail.

How might we proceed to test the manufacturer's claim? We should test the claim against the hypothesis that there is no difference (the status quo, or null hypothesis), which can be written as H_0: The distribution of X is the same as the distribution of Y. If there is really no difference between years, we should expect the Xs and Ys to be fairly well mixed in a combined ranking of the data. If the claim is justified, we should expect the Ys to be generally larger than the Xs. For the given data, the combined ranking is

$$17.4(X), \ 18.6(X), \ 20.2(Y), \ 21.0(X), \ 21.9(Y), \ 23.6(Y), \ 24.3(Y).$$

To obtain a suitable test statistic, we first replace the data values by their ranks. Let $R(X_i)$ be the rank of X_i in the combined ordering and let T be the sum of all $m = 3$ of the X ranks. By the reasoning of the previous paragraph, we should reject H_0 in favor of the manufacturer's claim if T is "small enough." Here $T = 1 + 2 + 4 = 7$, but to know if $T = 7$ is small enough, we need to know the distribution of T under H_0.

Let $N = m + n$. If the Xs and Ys do come from the same distribution, then all $(m + n)! = N!$ possible orderings (permutations) of the N data values are equally likely. To simplify calculations, notice that order *within each sample* is not important. Thus, the $m!n! = 3!4!$ orderings

$$X_1 < X_2 < Y_1 < X_3 < Y_2 < Y_3 < Y_4, \ldots, X_3 < X_2 < Y_4$$
$$< X_1 < Y_3 < Y_2 < Y_1$$

in our example all yield the same configuration $XXYXYYY$. There are $\binom{N}{m} = \binom{7}{3} = 35$ possible configurations, each of which is repeated $m!n! = 3!4!$ times for the different orderings. All orderings are equally likely, and so are all configurations. We can therefore calculate the distribution of T by considering only the (smaller) set of configurations. One configuration $(XXXYYYY)$ yields $T = 6$, one $(XXYXYYY)$ yields $T = 7$, and two $(XXYYXYY, XYXXYYY)$ yield $T = 8$. All thirty-five possibilities are listed in table 4.1.

From table 4.1, it is easy to get the sampling distribution of T. For example, $P(T=6) = P(T=7) = 1/35$, $P(T=8) = 2/35$, and $P(T=9) = 3/35$. The complete distribution is given in table 4.2.

TABLE 4.1

Configuration	Value of T	Configuration	Value of T	Configuration	Value of T
$XXXYYYY$	$1 + 2 + 3 = 6$	$XYYYXXY$	$1 + 5 + 6 = 12$	$YXYYYXX$	$2 + 6 + 7 = 15$
$XXYXYYY$	$1 + 2 + 4 = 7$	$XYYYXYX$	$1 + 5 + 7 = 13$	$YYXXXYY$	$3 + 4 + 5 = 12$
$XXYYXYY$	$1 + 2 + 5 = 8$	$XYYYYXX$	$1 + 6 + 7 = 14$	$YYXXYXY$	$3 + 4 + 6 = 13$
$XXYYYXY$	$1 + 2 + 6 = 9$	$YXXXYYY$	$2 + 3 + 4 = 9$	$YYXXYYX$	$3 + 4 + 7 = 14$
$XXYYYYX$	$1 + 2 + 7 = 10$	$YXXYXYY$	$2 + 3 + 5 = 10$	$YYXYXXY$	$3 + 5 + 6 = 14$
$XYXXYYY$	$1 + 3 + 4 = 8$	$YXXYYXY$	$2 + 3 + 6 = 11$	$YYXYXYX$	$3 + 5 + 7 = 15$
$XYXYXYY$	$1 + 3 + 5 = 9$	$YXXYYYX$	$2 + 3 + 7 = 12$	$YYXYYXX$	$3 + 6 + 7 = 16$
$XYXYYXY$	$1 + 3 + 6 = 10$	$YXYXXYY$	$2 + 4 + 5 = 11$	$YYYXXXY$	$4 + 5 + 6 = 15$
$XYXYYYX$	$1 + 3 + 7 = 11$	$YXYXYXY$	$2 + 4 + 6 = 12$	$YYYXXYX$	$4 + 5 + 7 = 16$
$XYYXXYY$	$1 + 4 + 5 = 10$	$YXYXYYX$	$2 + 4 + 7 = 13$	$YYYXYXX$	$4 + 6 + 7 = 17$
$XYYXYXY$	$1 + 4 + 6 = 11$	$YXYYXXY$	$2 + 5 + 6 = 13$	$YYYYXXX$	$5 + 6 + 7 = 18$
$XYYXYYX$	$1 + 4 + 7 = 12$	$YXYYXYX$	$2 + 5 + 7 = 14$		

TABLE 4.2

k	6	7	8	9	10	11	12	13	14	15	16	17	18
$35P(T=k)$	1	1	2	3	4	4	5	4	4	3	2	1	1

Recall that $T = 7$ for our example, corresponding to the configuration $XXYXYYY$. When H_0 is true, there are only two chances in thirty-five of observing a value of T at least as small as 7. Since the event $T \leq 7$ is fairly improbable if X and Y have the same distribution, T is "small enough" for us to reasonably accept the manufacturer's claim as valid.

Suppose, instead, the manufacturer had claimed that current models get at least 10 percent better gas mileage than comparable models from last year. How can the procedure be modified to test this new claim? Since $\overline{X} = (17.4 + 21.0 + 18.6)/3 = 19$, we could write the hypothesis to be tested as H_0': The distribution of $X + 1.9$ is the same as the distribution of Y. This suggests that we add 1.9 to each X value and recompute T. Doing this yields the new data

$$X' = X + 1.9: \quad X_1' = 19.3, \quad X_2' = 22.9, \quad X_3' = 20.5.$$

The new combined ranking is

$$19.3(X'), 20.2(Y), 20.5(X'), 21.9(Y), 22.9(X'), 23.6(Y), 24.3(Y).$$

Since the Xs now occupy positions 1, 3, and 5, the new value of T is $T' = 1 + 3 + 5 = 9$.

From table 4.2, we see that there are seven chances in thirty-five of observing a value of T' at least as small as 9 when H_0' is true, and so T' does not provide very strong support for the claim of a 10 percent increase in gas mileage. From these limited data, we can reasonably conclude that new models get better mileage than those of last year, although the increase is not as great as 10 percent.

Conclusion

Our point of view here is that probability models provide a basis for drawing conclusions in situations that contain elements of uncertainty. Since models inevitably require us to make assumptions, we need to question repeatedly the validity of those we make. Furthermore, any conclusions we draw can be no more reliable than the data on which they are based. In short, to paraphrase the preface of [4], we want our students to develop a critical attitude toward the use of the models they select and to be cautious in the interpretation of their results. Graphical and informal techniques play a vital role in the development of this attitude. Therefore, they should be introduced early and used repeatedly and not allowed to sit in some introductory unit gathering dust from lack of use throughout the remainder of the course.

5. Graphically Speaking: Primary-Level Graphing Experiences

Laura Duncan Choate

JoAnn King Okey

I⊤ IS said a picture is worth a thousand words, and what is a graph but a picture? Kids and graphs are a great combination, but over the years we have failed to provide children with the numerous graphing experiences they need. Our primary mathematics textbooks have offered few pages on graphing. Daily, perfect graphing situations occurred, but we were unable to take advantage of them. It became so time-consuming to prepare class-sized materials each time we wanted to graph that we were forced to pass up ideal opportunities. What good is an idea if it can't be put to practical use?

We tried to think of logical ways to make graphing easy and practical enough to use on a regular basis and a structure to guide our students from concrete to abstract graphing experiences. Here are some solutions that were successful in our classrooms.

- Graph-it bags sequenced to develop concepts in graphing using concrete objects with two to five attributes
- A reusable class-sized grid that can easily be displayed whenever a graphing situation occurs
- A flexible series of pictograms developed for use with this grid covering a multitude of primary educational lessons
- Folders mounted with pictures and filled with correlating grids to provide independent, self-paced graphing practice
- A collection of graphing suggestions encompassing a child's situation in all academic areas, with particular emphasis on the composition of the class itself and the individual characteristics of the children in it

The more we graphed, the more impressed we were with the insight our

33

students developed, and when a graphing situation arose, we were prepared. When we shared our ideas with other teachers, they became as enthusiastic as we were about extending graphing activities throughout the year. Although graphing is not a new idea, we hope we can offer some new twists on the traditional methods and help others to realize the full potential graphing has to offer.

Why Graph?

Graphing is a visual presentation of statistics. It is simpler in concept and form than numerical presentation. Graphs provide a visual structure to order masses of figures and help us internalize the concept of numerical amounts. Graphing helps a child see that mathematical knowledge is an integral part of daily life and can be applied to many situations. As adults, we see graphs almost every day and are expected to synthesize information from them. Statistics compare values and allow logical conclusions to be drawn. Graphing is a useful tool not only for interpreting direct results but also for building inferential hypotheses.

Graphing gives the child an opportunity to compare, count, add, subtract, sequence, and classify data. Children need to experience mathematics firsthand to understand how it affects their daily lives and to draw mathematical conclusions about the world around them. A tactile and visual representation of amounts facilitates children's understanding of comparative values.

Developing the Theory of Graphing
with Children

Beginning graphers should experience the same situation sequentially through the developmental levels of graphing to help them comprehend the relationship between reality and its representation in an abstract graph. The repetition of stages, from tactile to symbolic, helps them internalize the developmental theory behind the concept of statistics. Here are three types of graphs, presented sequentially, that can bridge the gap from real to abstract.

Object graphs use actual things for the child to compare (see fig. 5.1). This level is most important because it forms the foundation for the comprehension of graphs.

Picture graphs use pictures of objects to stand for real things (see fig. 5.2). These are more abstract than object graphs because a picture only represents reality yet is visually related to the object. This level is important because picture graphs help span the gap between the object and the abstract symbol.

Fig. 5.1. Object graph

Fig. 5.2. Picture graph

Symbolic graphs use symbols to stand for real objects, such as the paper stickers in figure 5.3. This is the most abstract level, since the child must transfer symbols back to reality to find meaning. A gummed paper square on a graph can only represent, abstractly, a real piece of candy.

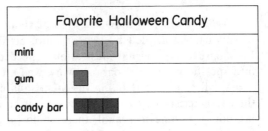

Fig. 5.3. Symbolic graph

Graphing Techniques

Group graphs

A common difficulty in today's busy classroom is the lack of preparation time for making the multitude of class-sized graphs needed to meet every appropriate graphing situation. Some minor inconveniences can become major obstacles in the classroom, where every second counts. Our solution was to develop reusable classroom grids for each graphing level.

Level 1: Object graphs and people graphs. A set of objects with two or more common characteristics can be divided by attribute and laid on the reusable grid (see fig. 5.4). A reusable grid can be made by ruling 8-cm squares on a 6.4 m × 9.6 m piece of tagboard and covering it with clear contact paper or laminating film for durability.

Divide a group of children according to two or three characteristics; graph with the children themselves (see fig. 5.5). A reusable floor grid can be made of 1 m × 5 m heavy plastic or oilcloth marked into 30-cm squares. Lines can be drawn with permanent pens or made with cloth tape.

Object graphs on reusable grids are obviously temporary. A record of such graphs can be maintained through photographs or transferring information to picture graphs. Permanent object graphs can be made by gluing or otherwise securing the objects to the grid.

Fig. 5.4

Fig. 5.5

Level 2: Picture graphs. A group of small pictures representing a set of objects can be divided by two or more attributes and attached to a grid. For nonreaders, a visual illustration is needed to define the attributes used in a specific graph. Pictograms work well on a reusable grid. A pictogram is a small picture that fits in the 8-cm square and can be used to illustrate the characteristics by which you are going to graph. For example, in graphing how we come to school, there could be pictograms of a bus, a car, a walker, and a bike (see fig. 5.6).

Pictograms are usually placed in the squares farthest to the left (to emphasize left-to-right sequence), the bottom, or the top of the grid. Pictograms can be sketches, photographs, or cut-outs from magazines. They may or may not be labeled with the word for the picture. The child's photo or self-portrait (representing the individual) or duplicates of the pictograms (representing the objects) are used to record the graphing experience.

Fig. 5.6

A reusable grid can be made in the same way as for an object graph. Such a graph is usually displayed against a wall or on a bulletin board for greater visibility.

Against a wall, paper-clip hangers can be used. Cut a small 1-cm slit in the middle of the top edge of each square. Insert a paper clip into each slit. Pictures can be secured under the edge of the clip. On a bulletin board, pins or tacks can be used to post and change pictures.

Children can record their graphing responses with individual portraits or photographs. Miniature photographs can be found on cumulative records. Small headshots can also be obtained by photographing three children together and cutting the photograph apart. In this way, one roll of ten-exposure film produces portraits for a class of thirty. Series of permanent graphs are often useful throughout the year, and photocopies of the photographs or self-portraits can be made.

Level 3: Symbolic graphs. Record information on a grid, using an abstract, symbolic form, as in figure 5.7. The same grid used for picture graphs is suitable for symbolic graphs. Attributes can be designated with pictograms or labels. Paper squares, beads on shoelaces, Unifix cubes, gummed squares or circles, blocks, lines, or bars can be used as symbols to record the graphing situation.

Fig. 5.7

Reading and Interpreting a Graph

No graphing experience is complete until questions are formulated and answered. Children should always be encouraged to discuss and interpret the graph in order to discover patterns and relationships.

Answering questions requires students to react to the graph, and eventually they become comfortable asking questions and discussing situations without the teacher's leadership. Questions to ask about the "How We Get to School" graph might include these: Which row has the most? The fewest? Are any rows the same? Are there more bike riders or bus riders? Are there more walkers or car riders? Are there fewer bike riders or fewer car riders? How many bike riders are there? How many walkers? How many more bus riders are there than walkers? How many fewer bike riders are there than car riders? How many people ride bikes or buses? How many people are there altogether?

Many of these simple statements can be translated into symbolic mathematical language. "There are six car riders and five walkers" becomes "$6 + 5 = 11$."

Individual Graphing Activities

The following individual activities can be self-paced to reinforce the graphing concepts introduced in the group graphs. They provide experience on both the object and the symbolic levels and are arranged in order from simplest to most complex.

Graph-it bags

Graph-it bags are easily developed, inexpensive aids that can make use of leftovers or junk; they provide not only graphing experiences but many tactile experiences as well. A small bag, lunch sack, zip-lock bag, or cloth bag with drawstrings is filled with eight to twelve objects that can be classified by a single attribute, such as size, shape, or color. Graph-it bags might include such items as buttons, coins, shells, edible nuts, ceramic tiles, bottle caps, rocks, old keys, beans, pasta, leaves, crayons, small toy animals, nuts and bolts, jar covers, old playing cards, toy soldiers, toy vehicles, checkers, seeds, seed pods, fruit pits, postcards, and nails.

The child determines some common attribute of the objects in a bag (this can differ from your original intent and still be valid) and arranges the objects by attribute, one object to each square, on a reusable object grid, as in figure 5.8.

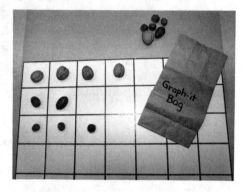

To complete the assignment, the child should verbalize (ideally to an adult or older tutor) three statements about what he or she sees in the graph:

Fig. 5.8

"There are more walnuts than pecans."
"There are fewer pecans than hazelnuts."
"There are two more walnuts than pecans."

Ideally, the teacher could develop a series of graph-it bags, beginning with two obvious attributes and extending to multiple, higher-level graphing experiences.

Graphing folders

Graphing folders provide experiences on a more abstract level (see fig. 5.9). Carefully choose an illustration, photograph, picture, or drawing that enables a child to count two or more types of objects easily. Avoid pictures that show only part of an object, such as half a banana. Mount the illustration on the front of a manila file folder. Fill the inside with dittoed copies of a teacher-made grid with appropriate attribute names or illustrations. Each child chooses a folder and takes out a copy of the grid. Looking at the illustration, the child counts the objects and records his or her findings on the grid.

Fig. 5.9

A series of individual folders can be ranked from simplest to most complex by grouping them by number of attributes, beginning with pictures having two attributes. Five to six folders at each attribute level will allow individuals to practice their graphing skills before extending their mastery to a higher level.

Both types of individual graphing experiences, the graph-it bags and graphing folders, should be preceded by successful experiences with class graphs, particularly by the use of real-object grids in a group situation, with questions, answers, and conclusions elicited by the teacher.

Seizing the Moment

Graphing is something you can do every day or with every unit of study. It is appropriate in any situation in which numbers, values, and amounts are open to comparison. After several planned graphing experiences, it should become easier to recognize the many natural opportunities that occur daily in the classroom.

Continuing graphs

Many graphs have more meaning when maintained throughout the school year, allowing for more data to be collected and compared. One such graph is shown in figure 5.10. Those graphs used daily in some classrooms might include the following:

Attendance (present/absent)	Which story do you want?
Temperature (indoor/outdoor)	Milk orders
Who lost a tooth?	Sight words
Weather	Activity center choices
Will you share today?	Will you play indoors or outdoors?
Lunch count	Progress charts

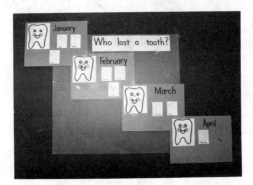

Fig. 5.10

Class characteristics

A common theme at the beginning of the school year is becoming acquainted with each other and developing a sense of group and individual worth. Learning about one's self and one's relationships to the group provides an ideal situation for group graphs (see fig. 5.11). Appropriate topics might be boys/girls, birthdays, ages, hair color, eye color, pets, bedtimes, color of clothes, height, weight, right- or left-handedness, number in family, glasses or no glasses, first letter of names, fingerprint type (whorl/arch/loop), transportation to school, birthplace, last year's teacher, and lunch items.

Fig. 5.11

Graphs on particular academic subjects

Reading—Beginning sounds, vowels in a paragraph, consonant letter frequency, letter order (names, words)

Math—Height of children, plants, toys, etc.; weight of children, blocks, etc.; hand span; area of desks, papers, etc.; shoe size; length of arms, legs, etc.; weights of objects (in number of blocks)

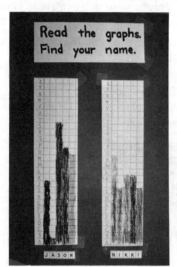

Science—Leaf perimeter, weekly plant growth, types of homes animals live in, magnetic or not, weather, birds seen, whether objects will sink or float, whether or not you hear the sea in the shell

Music—Favorite song, types of notes in a tune, tempo of tunes, happy or sad music, what the music reminds you of

Physical education—Number of times you can jump rope, number of times you can bounce a ball, how high you can jump, how fast you can run, league win/loss record

Our favorites

Children love to express their opinions, and preference polls are an ideal way to find out everyone's opinion. Topics for this type of graph could include each person's favorite fruit, vegetable, snack, holiday, color, TV show, pet, candy bar, fairy-tale character, ice cream flavor, zoo animal, dinosaur, classroom job, lucky number, type of apple (sauce/raw/juice), and school activity.

We have suggested only a few of the many opportunities for graphing. The more frequently graphing is incorporated with the class, the more graphing opportunities will become apparent.

6. Using Teaching Devices for Statistics and Probability with Primary Children

Kathleen A. Knowler

Lloyd A. Knowler

Aт тне kindergarten level most children are learning basic mathematical terminology and concepts—shapes, numerical recognition, classifying, comparing, and counting. In conjunction with this, they can begin to learn some descriptive statistical techniques and elementary probability concepts.

Pictorial graphing is frequently used at this level. Initially, concrete objects are used directly. For instance, children can keep track of who is in school by putting a card in a slot to indicate their presence (see fig. 6.1). Cards with their own pictures on them are fun (or clothespins with their names). Boys and girls make an easy distribution so that totals are not too large and the two groups can be compared. It is also possible to make non-sexist distinctions by dividing the chart by bus riders and walkers, or five-year-olds and six-year-olds, or any other division that makes sense.

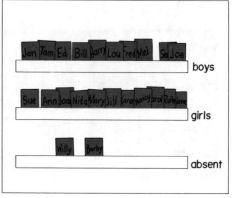

Fig. 6.1

Other graphing ideas also can involve the children personally; eye color, hair color, pets, and birthdays are all interesting subjects. These things can be graphed by having the children stand in groups according to eye color, for example, and then paste drawings of their eyes on paper—brown on one paper, blue on another.

Sorting activities are frequently used in kindergarten for building sets, counting, and comparing. The grab game is fun. Children are given a bag containing a number of blocks of two or three shapes (or colors). They draw a block in a random manner and group it with other similar blocks they have drawn. When they have finished, they can count the number of blocks of each shape in their sampling. The number of shapes can be varied to produce different distributions, and the children can be asked to predict which will be drawn. It is probably best not to use replacement at first so the children can have concrete evidence of what they drew in front of them to be counted and compared. At the older age levels, however, they can keep a written or pictorial record of what is drawn.

Another form of this game is for two blocks to be drawn simultaneously and records kept of how often two of the same shape appeared.

In first grade, extensions can be added to the same activities. At this point addition and subtraction are being introduced. Many games originally designed to practice number facts can also be used to provide statistical or probability data.

A very popular game at this level is two-dice adding. Two dice are thrown, their sum is calculated, and the appropriate square on a graph is colored in to indicate the sum (fig. 6.2). Six-year-olds generally are not ready for competitive games, but this can be played by two children side by side, the object being to see what sum wins on each child's graph.

This game can also be played by a large group using giant dice made from milk cartons. (These "dice" are often biased, depending on how care-fully they are made, but this leads to interesting comparisons.) Older chil-dren who play the adding game can be given loaded dice to see how the results differ. (Will they notice the difference without knowing that the dice are loaded?) Other extensions include using more than two dice, or adding or subtracting a third die to or from the result. In all cases, after sufficient trials the pupils should observe the distribution that results.

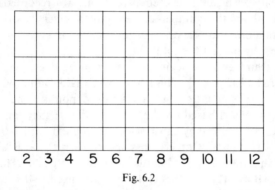

Fig. 6.2

First graders begin to learn about measuring. A good introduction to the study of time is an egg timer. The children can be asked to predict how often they can do a certain task (write their name, hop across the room, etc.) in three minutes. Generally they like to do this repeatedly, and so their scores can be saved and a record made. (Older children can find averages—arithmetic means—either of their own repeated trials or by comparing classmates' trials.)

Another timer that can be used is a stopwatch. Again, records can be kept (e.g., how long it takes to run across the playground) and averages computed. Some ideas about variation may begin to emerge.

Pupils at this level also begin to learn linear measuring. Tracings of hands or feet provide a concrete representation for a graph of body sizes. Again, these can be ordered and the distribution studied. Older children could record measurements, and correlations could be examined. Can foot size be predicted from hand size?

Although first graders have some difficulty with competitive games, these can be played with an adult participating. Games involving some physical task are usually most popular. A ring-toss game can be made from a soft-drink bottle and the rims cut from margarine tub lids. Children toss five or more rings and keep score over several trials. Comparisons can be made based on the positions from which the rings are thrown or on repetitions. A simple bowling game can be played with a soft, spongy ball and milk cartons as pins. Similar statistics from this game can be studied.

At holiday times many good estimation and prediction problems can be developed from special treats. At Halloween, minisized bags of different-colored candies can be purchased. One bag can be opened for a group to sort and count according to color. Then each pupil can be given his or her own bag. At first, predictions can be made about the number of candies of each color and the total number in each bag. Children can then find the results for their own bags. Beyond first grade, graphs can be drawn to show these results.

At Easter, estimating jelly beans is popular. Each child guesses the number of beans in a jar, and their predictions are recorded. Guesses can also be made about the number of beans of a given color. Older pupils could draw random samples to see how many of each color are in the sample before they predict the total.

Second and third graders begin to enjoy board games. One such game is "Hard-to-Hardest" (fig. 6.3). This game involves deciding whether to take a chance on harder questions to score more points. The players move markers along a board based on the roll of a die. If they land on a square marked "Take a Card," they must select a problem card labeled "Hard," "Harder," or "Hardest." Correct answers to "Hard" cards allow a player to move two squares; "Harder" cards, four squares; and "Hardest" cards, six squares. After playing, the children can discuss the probability of winning or losing, depending on the choices made.

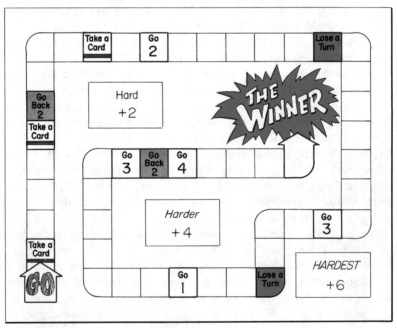

Fig. 6.3. Hard-to-Hardest

As pupils become more familiar with numbers and some basic concepts of probability, they are ready to explore new ideas.

7. Statistical Sampling and Popsicle Sticks

Jerry Shryock

CLASSROOM activities can easily be planned to enable children as early as second or third grade to understand the significance of a random sample taken from a large population. One such project might involve a large quantity of Popsicle sticks.

Illustrative Lesson on Sampling

Harry Reynolds, fourth-grade teacher, prepared for the project by marking 400 Popsicle sticks with colored dots. Using felt-tip pens, he marked 100 with a red dot, 200 with blue, and 100 with yellow. Then he thoroughly mixed all 400 of the sticks in a large grocery bag. The next day in class Reynolds held up the bag and told his class that a mixture of 400 Popsicle sticks marked in red, blue, and yellow were in the sack. He then posed this question: "How can you find out if there are more sticks marked with a blue dot than with red or yellow ones without actually counting all 400?"

After several minutes of class discussion Mr. Reynolds suggested that if just several handfuls of the sticks were taken from the bag and sorted, the class could perhaps tell whether more sticks of one kind than another were in the bag. He also noted in further classroom exploration that if a subset of a larger population is chosen in this manner, then the subset is called a *sample.*

Ellen said, "We'll have to be sure to reach into the sack with our eyes closed, so we get a sample just by chance. Otherwise, the person choosing the sticks might have a favorite color and pick more sticks of that kind than any other."

Additional comments focused on this importance of getting a sample in a way that avoids any bias. On checking a resource book, the class discovered that their sample could be called a *random sample* only if every stick in the sack had the same chance of being chosen. With this information the children had to be assured that the sticks had been mixed together thoroughly so that a handful from the bottom of the sack would not give results different from a handful from the middle.

Next the teacher asked, "What do you think would be a reasonable size for our random sample? Remember that we want a sample that will stand for, or represent, the population and will give us a good idea of how many of the three different kinds of sticks are in the bag."

Johnny answered, "If we take a sample of just six or seven Popsicle sticks, they might all be the same color. If we take 300 or more sticks for our sample, we haven't saved much time over counting all 400. I think a sample size of 125 would be about right."

Mr. Reynolds replied, "Suppose we follow Johnny's suggestion for a sample size and ask different class members to close their eyes, reach into the bag, and take out handfuls of sticks until we have 125 Popsicle sticks in our sample."

As different students reached into the sack and removed a handful of sticks, other class members separated the sticks into three stacks according to their dots of color. Another student at the chalkboard recorded the results in the form of a table (fig. 7.1).

Information Table

Popsicle Stick Sample of 125		
Kinds of Sticks	Count	Totals
Marked in Red	ЖТ ЖТ ЖТ ЖТ ЖТ ЖТ II	32
Marked in Blue	ЖТ ЖТ ЖТ ЖТ ЖТ ЖТ ЖТ ЖТ ЖТ ЖТ ЖТ ЖТ III	63
Marked in Yellow	ЖТ ЖТ ЖТ ЖТ ЖТ ЖТ	30

Fig. 7.1

Next Mr. Reynolds suggested that besides using the results of their sample to make a table, a model graph could be made by placing the sticks side by side in three rows according to color. After several minutes of spirited discussion on how best to make a model graph, several class members were assigned the task of arranging the sticks in rows on a long sheet of butcher paper placed on the floor of the classroom (fig. 7.2).

After the graph was completed, the children commented that the graph showed more clearly than the table what the results of the sample were.

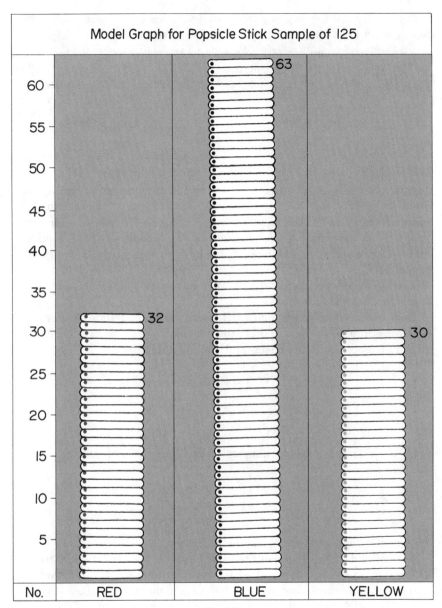

Fig. 7.2

Linda: "If our sample is representative of the population of 400, it looks as if there should be about twice as many blue sticks in the sack as red ones and twice as many as yellow ones."

Teacher: "That seems to be a good conclusion, Linda. Can you tell us how you got that estimate?"

Linda: "By just looking at our model graph, it's easy to see that the blue row is about twice as tall as the red and yellow rows."

Rick: "There's another way to get that same estimate. I divided 63, the number of blue sticks, by 32 and then by 30, the number of red and yellow ones. I got close to 2 in each case." He put his work on the chalkboard:

$$
\begin{array}{c}
32 \overline{)\,63} \\
\underline{32} \\
31
\end{array}
= 1\,\frac{31}{32} \text{ (rounds to 2)}
\qquad
\begin{array}{c}
30 \overline{)\,63} \\
\underline{60} \\
3
\end{array}
= 2\,\frac{3}{30} \text{ (rounds to 2)}
$$

After spending a few minutes with the class considering the answers given by Linda and Rick, the teacher divided the students into four-member teams and asked each team to estimate the answer to this question: "Using the results of the random sample we took, what is a good estimate of the number of red, blue, and yellow Popsicle sticks in the population of 400?"

The next day the teacher asked what estimates had been made by the teams and for what reasons.

Donny said, "Our team thinks that there are around 100 red sticks, 200 blue ones, and 100 yellow. To get these estimates we first divided 400, the size of the population in the bag, by 125, the sample size:

$$
\begin{array}{c}
125 \overline{)\,400} \\
\underline{375} \\
25
\end{array}
= 3\,\frac{25}{125} \text{ (which rounds to 3)}
$$

Then we just multiplied the number of each color of Popsicle stick in the sample by 3:

$$
\begin{aligned}
3 \times 32 &= 96 \text{ (red)} \\
3 \times 63 &= 189 \text{ (blue)} \\
3 \times 30 &= 90 \text{ (yellow)}
\end{aligned}
$$

We rounded these products to get our estimate."

Jeri Lu said, "Our team got the same answers, but we did our work in a different way. We let one part of the sample be red sticks and then two parts should be blue, because in the sample there were about twice as many blue sticks as red ones. There should also be one part of the sample yellow sticks, because we drew out of the bag about as many yellow sticks as red ones.

"Since 1 + 2 + 1 is 4, we think that the sack of 400 Popsicle sticks has four parts. Dividing 400 into four parts gives 100 for the size of each part."

She then wrote on the chalkboard:

$1 \times 100 = 100$, so 100 red sticks

$2 \times 100 = 200$, so 200 blue sticks

$1 \times 100 = 100$, so 100 yellow sticks

Comments from excited students about these methods of solution convinced Mr. Reynolds that this was an excellent place to introduce concepts of ratio and proportion.

After helping the children explore these two topics, he asked the teams what procedures they would use to estimate the number of left-handed students in their school building. Questions to be considered were: "How can your team get a sample of the students that is without any bias?" "Is your random sample representative of the population?" "What size should the sample be?" "Will it help to prepare a table and a graph of the results of the sample?" "How will you use proportions in the sample to estimate how many left-handed students are in the building?"

8. Fair Games, Unfair Games

George W. Bright

John G. Harvey

Margariete Montague Wheeler

CHILDREN and adults sometimes say, "But that's not fair!" This may mean, "Things aren't going my way and I want them to," but it may also signal that a bias has been identified. This article will examine mathematical fairness and unfairness as it can be developed using games in intermediate and middle grade classrooms. Games are used because students eagerly play games. Psychologically and socially, games are easily separated from real-life activity, so they provide a neutral, nonthreatening context in which fairness can be examined.

Fairness can be described indirectly by calling attention to unfairness: a situation is seen as fair if no one involved in it has an obvious advantage; that is, if the game is not unfair. Fairness can also be dealt with directly:

49

for example, a simple situation, readily accepted as fair, is flipping an ordinary coin. Either heads or tails can land up, and most people accept as a given that if many tosses are made, each side will land up about the same number of times. That is, the situation is "fair."

If we alter the situation so that both sides of the coin show a head, the situation becomes grossly unfair. Saying that either a head or a tail can land up is not realistic, since only one of these outcomes can happen. That is, the situation is "not fair."

Helping students recognize when a situation is fair or unfair is a reasonable expectation of the school curriculum. The games that will be presented in this chapter are examples of activities that can be used to teach the distinction.

In complex situations it may be difficult to determine mathematically whether a situation is fair. Sometimes correctly applying the appropriate mathematics requires considerable experience and insight. Elementary and junior high school students can begin, however, to develop an understanding of fairness by gathering data.

In the games that follow, the purpose of collecting data about winners and losers is to determine whether each player has the same opportunity to win. On the basis of the data collected, students can begin an analysis of whether the game is fair.

Collecting data for a game is simple: play the game repeatedly and keep a tally of which player wins most often. An analysis of whether a game is fair, however, is more difficult. On the basis of data, one can determine only whether the game *seems* to be fair, since data usually contain random fluctuations, called "noise." Suppose Sam and Anne play a game and Anne wins eleven games out of twenty. It is difficult to say with confidence that the game is unfair. Of course it may be. But suppose Sam wins eighteen times out of twenty. Now one is more likely to conclude that the game is unfair. Such a conclusion cannot be stated with certainty, however, because Sam's apparent advantage may only reflect "noise" in the data. In any data-gathering situation, rare events may occur, and they *may* occur more than once. Thus one can never be absolutely sure of conclusions based on data and can only hope to be confident about those conclusions.

Conclusions based on data may be in error for a variety of reasons. In games that involve chip flipping, discrepancies might be caused by unbalanced chips. That is, one of the two sides has a greater probability of landing up. When dice are rolled, the dice might not be shaken adequately, or they might be "loaded." If balls are pulled from urns, the balls might not be thoroughly mixed. When a spinner is used, friction might cause it to stop in one particular position more often than expected. These kinds of problems usually cause systematic discrepancies that can be detected by close examination of the data. Such studies are both instructive and interesting, but they are not our focus.

Chips off the Old Block

Game 1   (2 players)

You will need

2 red chips with an A side and a B side

1 blue chip with an A side and a B side

paper and pencil to keep score

Game rules

1. Decide which player will be 1 and which player will be 2.
2. Flip all three chips at the same time. Player 2 scores a point if both red chips show A or the blue chip shows A or both. Otherwise, player 1 scores a point.
3. Play 16 rounds.
4. The winner is the player with more points at the end of 16 rounds.
5. Play two or three games. Then answer these questions:
   - Does each player have an equal chance at winning?
   - Does the same person win each time?
   - Is the game a fair game?

Game 2   (2 players)

You will need

3 yellow chips with an A side and a B side

1 green chip with an A side and a B side

paper and pencil to keep score

Game rules

1. Decide which player will be 1 and which player will be 2.
2. Flip all four chips at the same time. Player 1 scores a point if all three yellow chips show A or the green chip shows A or both. Otherwise, player 2 scores a point.
3. Play 24 rounds.
4. The winner is the player with more points at the end of 24 rounds.
5. Play two or three games. Then answer these questions:
   - Does each player have an equal chance at winning?
   - Does the same person win each time?
   - Is the game a fair game?

Diet Fractions

Game 1   (2 players)

You will need

2 standard numbered dice
paper and pencil to keep score

Game rules

1. Decide which player will be A and which player will be B.
2. Roll the dice at the same time. Use the two numbers to make a fraction less than or equal to 1.
3. If the fraction is not reduced to lowest terms, player A scores a point. Otherwise, player B scores a point.
4. Play 12 rounds.
5. The winner is the player with more points at the end of 12 rounds.
6. Play two or three games and then answer these questions:
   - Does each player have an equal chance at winning?
   - Does the same person win each time?
   - Is the game a fair game?

Game 2   (2 players)

You will need

1 orange standard numbered die
1 blue standard numbered die
paper and pencil to keep score

Game rules

1. Decide which player will be A and which player will be B.
2. Roll the two dice at the same time. Make a fraction with the number on the orange die as the numerator and the number on the blue die as the denominator.
3. If the fraction is greater than 1, player A scores a point. If the fraction is less than 1, player B scores a point. If the fraction equals 1, each player scores a point.
4. Play 12 rounds.
5. The winner is the player with more points at the end of 12 rounds.
6. Play two or three games and then answer these questions:
   - Does each player have an equal chance at winning?
   - Does the same person win each time?
   - Is the game a fair game?

Find the Fairer Game

Play three rounds of each game below. Keep track of who wins each game, the odd or the even person. When you finish, answer the following questions.

1. Which game is fairer?
2. Why is the one game fairer than the other? (Make a table to help you.)

Game 1   (2 players)

You will need

2 standard numbered dice
paper and pencil to keep score
clock for timing

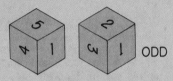

ODD

Game rules

1. Choose one person to be "odd" and the other person to be "even."
2. Roll the dice and find the difference between the two numbers.
3. If the difference is an odd number, the "odd" person scores a point. If the difference is an even number, the "even" person scores a point. Remember, 0 is an even number.

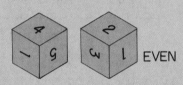

EVEN

4. Roll the dice for 2 minutes.
5. The winner is the person with more points.

Game 2   (2 players)

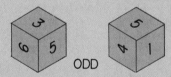

ODD

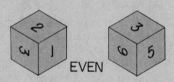

EVEN

You will need

2 standard numbered dice
paper and pencil to keep score
clock for timing

Game rules

1. Choose one person to be "odd" and the other person to be "even."
2. Roll the dice and find the product of the two numbers.
3. If the product is an odd number, the "odd" person scores a point. If the product is an even number, the "even" person scores a point.
4. Roll the dice for 2 minutes.
5. The winner is the person with more points.

Reprinted with permission from *Developing Mathematical Processes*, Level 6, Topic 85. © 1976 by The Board of Regents of the University of Wisconsin System for the Wisconsin Research and Development Center for Cognitive Learning.

Flipped Chip

Game 1 (2 players)

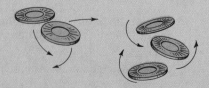

You will need

1 chip with an A side and a B side
15 blank chips for each player
1 gameboard for each player
paper and pencil to keep score

Game rules

1. Decide which player will be A and which will be B.
2. Flip the marked chip. If A comes up, player A places a blank chip on her or his car. If B comes up, player B puts a blank chip on her or his car.
3. The winner is the first person to fill all 15 chip spaces.
4. Play two or three games. Then answer these questions:
   - Does each player have an equal chance at winning?
   - Does the same person win each time?
   - Is the game a fair game?

Game 2 (4 players)

You will need

3 chips with an A side and a B side
15 blank chips for each player
1 gameboard for each player
paper and pencil to keep score

Game rules

1. Decide which player will be 3, which will be 2, which will be 1, and which will be 0.
2. Flip all 3 chips at the same time. Player 3 puts a chip on her or his car if three A's come up. Player 2 puts a chip on her or his car if exactly two A's come up. Player 1 puts a chip on her or his car if exactly one A comes up. Player 0 puts a chip on her or his car if no A's come up.
3. The winner is the first player to cover her or his car.
4. Play two or three games. Then answer these questions:
   - Does each player have an equal chance at winning?
   - Does the same person win each time?
   - Is the game a fair game?

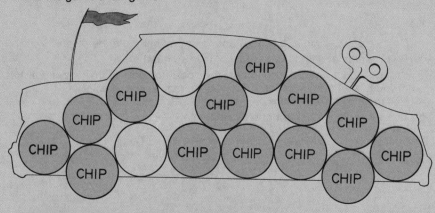

Iced Dice

Game 1   (2 players)

You will need

3 standard numbered dice
paper and pencil to keep score

Game rules

1. Decide which player will be A and which player will be B.
2. Roll the three dice and find the sum of the three numbers.
3. If the sum is even, player A scores a point. If the sum is odd, player B scores a point.
4. Play 20 rounds.
5. The winner is the player with more points at the end of 20 rounds.
6. Play two or three games and then answer these questions:
   - Does each player have an equal chance at winning?
   - Does the same person win each time?
   - Is the game a fair game?

Game 2   (2 players)

You will need

2 blue standard numbered dice
1 orange standard numbered die
paper and pencil to keep score

Game rules

1. Decide which player will be A and which player will be B.
2. Roll all three dice at the same time. Find the sum of the two numbers on the blue dice. Find the difference between that answer and the number on the orange die.
3. If the difference is odd, player A scores a point. If the difference is even, player B scores a point. Remember that zero is even.
4. Play 20 rounds.
5. The winner is the player with more points at the end of 20 rounds.
6. Play two or three games and then answer these questions:
   - Does each player have an equal chance at winning?
   - Does the same person win each time?
   - Is the game a fair game?

Matchbook

Game 1 (2 players)

You will need

1 chip with the letter X on both sides

1 chip with an X side and a Y side

paper and pencil to keep score

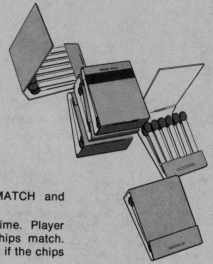

Game rules

1. Decide which player will be MATCH and which will be NO MATCH.
2. Flip both chips at the same time. Player MATCH scores a point if the chips match. Player NO MATCH scores a point if the chips do not match.
3. The winner is the first player to get 15 points. Play two or three games. Then answer these questions:
   - Does each player have an equal chance at winning?
   - Does the same person win each time?
   - Is the game a fair game?

Game 2 (2 players)

You will need

1 chip with an A side and a B side
1 chip with an A side and a C side
1 chip with a B side and a C side
paper and pencil to keep score

Game rules

1. Decide which player is MATCH and which is NO MATCH.
2. Flip all three chips at the same time. Player MATCH scores a point if two of the chips match. Player NO MATCH scores a point if all three chips are different.
3. The winner is the first player to score 15 points. Play two or three games. Then answer these questions:
   - Does each player have an equal chance at winning?
   - Does the same person win each time?
   - Is the game a fair game?

Number Spin

Game 1   (4 players)

You will need

1 spinner
paper and pencil to keep score

Game rules

1. The player whose first name begins with a letter closest to A is number 1. The next closest is number 2, and so on.
2. You will spin the spinner 50 times. Each time the spinner lands on a player's number, that player will get a point.
3. The winner will be the player with the most points.
4. Decide before playing who you think will win the game and why. Then play to see what happens.

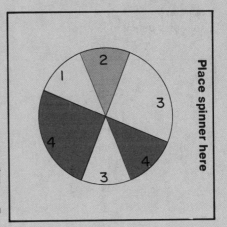

Place spinner here

Abadaca

Game 2   (4 players)

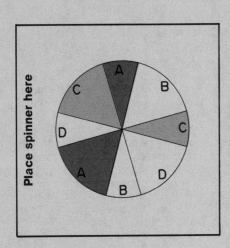

Place spinner here

You will need

1 spinner
paper and pencil to keep score

Game rules

1. Each player chooses one of the letters, A, B, C, D.
2. You will spin the spinner 50 times. Each time the spinner lands on a player's letter, that player will get a point.
3. The winner will be the player with the most points.
4. Decide before playing who you think will win the game and why. Then play to see what happens.

Spin Around

Game 1   (2 players)

You will need

1 spinner

paper and pencil to keep score

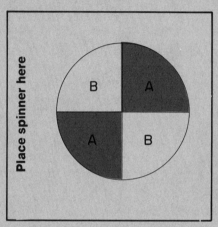

Game rules

1. The player whose first name has the most letters will be A. The other player will be B. If both names have the same number of letters, use last names. If this is still a tie, use middle names.

2. Spin the spinner for this game twice. Player A scores a point if the spinner lands on the same letter twice. Player B scores a point if the spinner lands on different letters.

3. The winner is the player with more points at the end of 20 rounds. Play two or three games. Then answer these questions:

   ● Does each player have an equal chance at winning?

   ● Does the same person win each time?

   ● Is the game a fair game?

Game 2   (2 players)

You will need

1 spinner

paper and pencil to keep score

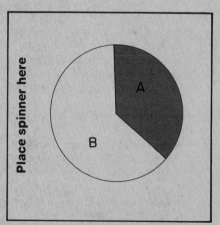

Game rules

1. Decide who will be player A and who will be player B.

2. Spin the spinner for this game twice. Player A scores a point if the spinner lands on the same letter twice. Player B scores a point if the spinner lands on different letters.

3. The winner is the player with more points at the end of 20 rounds. Play two or three games. Then answer these questions:

   ● Does each player have an equal chance at winning?

   ● Does the same person win each time?

   ● Is the game a fair game?

We want our students to be able to determine when a game seems to be fair. Making these decisions is a good first step in understanding basic concepts in probability. Although the games that follow can be analyzed mathematically, we encourage the reader to play each pair of games repeatedly, to gather data about winners and losers, and to make a personal determination about whether the games seem to be fair.

The eight pairs of games were designed to introduce fairness to students in elementary school, middle school, and junior high school. Each of the combinations of fair and unfair games (fair/fair, fair/unfair, and unfair/unfair) is represented at least once among the eight pairs.

Besides the gameboards, three kinds of equipment are needed—chips, dice, and spinners. The chips can be plastic chips like those used in many commercial games. If these are not available, cardboard disks or coins could be used. The dice are often standard dice. Blank dice can be labeled with gummed labels or with a grease pencil. Any two colors of dice can be substituted for the orange and blue dice called for; just be sure the students know the correspondence. The spinner can be a transparent one that fits over the gameboard. If not available, spinners can be made (see [92]).

All eight pairs of games were used in an experimental study of achievement grouping as it relates to the use of games in teaching mathematics [9]. The generating idea for all the games was the set of three games— Find the Fairer Game, Flipped Chip, and Number Spin/Abadaca—from *Developing Mathematical Processes* [105]. The students were eighty-two seventh graders (four classes taught by two teachers) in a Chicago suburb. Each pair of games was played for twenty minutes; game-playing sessions were held twice weekly for four weeks. A pretest and posttest of twenty-eight pairs of fair-unfair situations were given; students were asked to choose the fairer situation. The students were grouped with classmates of either similar or dissimilar achievement on the basis of pretest scores. The groups did not change during the four weeks.

Students' performance from pretest to posttest improved significantly. In the similar achievement group, the mean score increased from 15.7 to 16.9 and, in the dissimilar achievement group, from 16.0 to 17.7. Each increase was significant ($p < .01$), but a comparison between groups suggests that grouping did not affect learning.

The games were played without any explicit instruction by the teachers on fairness and unfairness. Whatever the students learned was from playing the games. This suggests that some games can be used effectively before formal instruction and that games do not have to be used only for drill.

9. Developing Some Statistical Concepts in the Elementary School

Harry Bohan

Edith J. Moreland

THE word *average* is part of the listening vocabulary of many children even before they enter school. Such phrases as "average rainfall," "average height," and "average income" are often used by teachers and textbooks alike before children comprehend the word.

The most common use of the term *average* comes in connection with our grading system. Statements like "Your spelling average is 70" are used by many teachers with pupils long before the children have command of the mathematical skills necessary to compute such an average. It is the act of computing the average of a set of numbers that gives most pupils an understanding of this concept. A procedure for developing the concept of average without computation is presented below.

The Average (or Mean)

On returning a set of papers to the class, a teacher gives each pupil a piece of adding machine tape equal in length (in centimeters) to his or her score. A pupil making a score of 80 will receive a tape 80 cm in length. Pupils can then get a physical representation of their scores by comparing their tapes with a meterstick.

If one follows the traditional 60–69, 70–79, 80–89, and 90–100 scheme for determining D, C, B, and A, respectively, a special meterstick might be marked as shown in figure 9.1 and used for making such comparisons.

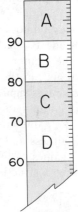

Pupils can compare their individual scores by comparing the lengths of their tapes. Such a practice can help develop the concepts of *greater than, less than,* or *equal to.*

Suppose Jeanette, who made 80 on her first test, made 60 on her second. With each test she receives a piece of tape, the first 80 cm in length and the second, 60 cm. She attaches the tapes end to end, and since she has had two scores, she folds the paper into two equal parts. Comparing the folded paper to the meterstick, she finds it to be 70 cm in length. This is her average. When the third test

Fig. 9.1

is returned, the tape representing that score is attached to the other two. This representation of the sum of the three scores is then folded into three parts of equal size and compared to the meterstick as before to determine her new average. The procedure can be followed as long as feasible.

Depending on the individual pupil or the makeup of the class, the teacher might, after each test, attempt to get a verbalization of what is meant by an average or a mean. Various situations might be presented from which pupils would try to predict what would happen to the average (or mean) if a test score was made that was higher than, lower than, or equal to that average. Predictions would be checked out by using adding machine tapes.

A similar activity involves dividing the class into groups of eight and determining the average height of each group. The procedure would be the same as in determining pupils' grade averages: attaching end to end the eight tapes representing the height of each student, folding the combined tape into eight parts of equal size, and measuring the folded tape to ascertain the average height.

An understanding of the mean (average) can also be promoted by a related activity making use of the overhead projector. A bar graph (as shown in fig. 9.2) is presented on the overhead as a month-by-month record of how last year's fourth-grade class raised money for the picnic at the end of the school year.

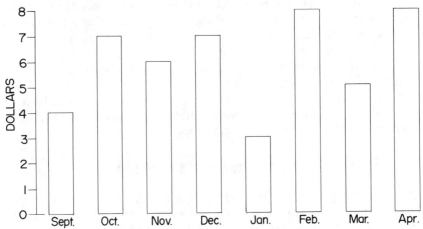

Fig. 9.2. Graph of monthly class savings

After they observe the graph, the children are asked, "How much will we have to raise each month for eight months to match what was raised by last year's class?"

We would now cut out strips of paper equal in length to each of the bars on the graph. The eight strips of paper (representing the total amount of money raised) would be taped together and folded into eight parts of equal length. The length of the folded paper represents the amount of money the class must raise each month for eight months to match last year's total.

To get an even clearer picture of the mean, eight colored strips of acetate, equal in length to the folded paper above, are cut out and placed on
each of the bars of the original graph on the overhead as in figure 9.3.

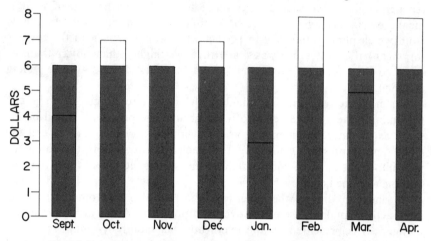

Fig. 9.3. Colored bars showing the average superimposed over original graph

The colored acetate represents the average amount of money raised each
month. The class can now be asked such questions as "Can you name the
months when the class was above average in raising money?" "Below
average?" "Were there any months when the class was right on the
average?" It might be a good idea to show another graph on which no
month was right on the average. Marking the graph with a scale of one
centimeter equal to one dollar would enable pupils to determine both the
total amount of money raised and the monthly average by comparing the
appropriate strips of paper to the meterstick.

The Median and the Mode

Although the terms *median* and *mode* are probably not in the vocabulary of most elementary pupils, these concepts are easily taught at this
level using little or no mathematics. It is worthwhile to demonstrate the
need for other measures of central tendency by pointing out the main weakness of the mean—the extent to which its value can be affected by extreme
scores.

This can be accomplished within the framework of activities discussed
earlier. In the activity concerning average height, make up the set of eight
people by taking seven of your students and including Bill Walton (the
basketball player), or someone's new baby sister, or some locally prominent
tall or short person as the eighth member. Show how this mean differs
from that of eight students in the class.

In the activity concerning averaging grades, show the effect that a grade of zero can have on the average (mean) of a student who has had three other grades with an average of 80. The median and mode can now be presented as different measures of central tendency that minimize the effect of extreme scores.

To get at the meaning of the terms *median* and *mode,* have pupils write their test score on an index card and drop it in a box. For our first try we want to use an odd number of scores. A card is pulled from the box at random and placed on a chart. A second card is then extracted, and the question is asked, "Is this number greater than, less than, or equal to the first number?" This second card is placed on the chart to the right of, the left of, or above the first card depending on whether the number on it is respectively greater than, less than, or equal to the number on the first card.

We continue to pull out cards, asking the same question over and over until all cards are arranged on the chart in order, left to right, from smaller to larger. Looking at all the cards on the chart, ask, "What is the score that was made by the most people?" This is, of course, the tallest column. *Mode* is the name given to the number that makes up the tallest column.

"Now let's find the score that is right in the middle of this set of scores." One way of doing this is to remove the highest and lowest scores from the chart simultaneously, one with each hand. This process is continued until only a single card remains. The number on this card is the middle score, and its mathematical name is *median.*

"Would we *always* eventually get down to a point where a single card remained?" After getting a consensus that we would not, discuss the conditions under which this would or would not happen, providing an opportunity to review the concept of even and odd numbers. Now place an even number of cards in the box, place them on the chart in order as indicated above, and eliminate cards simultaneously from either end until only two cards remain. Now try to get the class to agree that the best solution would be to call the point halfway between the two remaining cards the median.

In Conclusion

We have seen that the concepts of *mean, median,* and *mode* can be meaningfully taught at the elementary school level. Some might suggest that just because we *can* teach these concepts at this level does not necessarily mean that we *should* teach them. I suggest that the idea of an average is often used in the elementary school and is in much need of clarification. Although the concepts of mode and median are not generally thought of as elementary school topics, they can be used as a vehicle for applying several mathematical topics that *are* a part of the curriculum (for instance, metrics, greater than, less than, and equality)—especially if they are presented as suggested here.

10. Triples: A Game to Introduce the Facts of Chance

Eris Bailey

IN CERTAIN games of chance, such as roulette and poker, the probability that a particular course of action will give the desired outcome can be calculated. The need for evaluating the chances of events having particular outcomes gave rise to the theories of elementary probability. This same need can be used to introduce probability to students at the middle school level. The game of triples is a perfect vehicle to demonstrate this need.

The only materials required to play triples are four dice and three-by-three arrays of blanks or squares like those for tic-tac-toe. The arrays are to be completed with numbers that are determined by the throw of one, two, three, or four dice. The number determined by each throw is the sum of the dots on the tops of the tossed dice. Each throw must be recorded in an array, and each player or team completes its own array. The object of the game is to create ordered triples of number facts in rows, columns, or diagonals. However, the rows, or "across triples," must be addition facts; the columns, or "down triples," must be subtraction facts; and the "diagonal triples" must be multiplication facts. See figure 10.1.

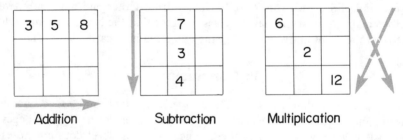

| Addition | Subtraction | Multiplication |

Fig. 10.1

Each player has two important choices to make during a turn—first, how many dice to roll, and second, where to place the resultant number. Both choices will decidedly affect the play of the game. Students soon begin to discuss the best number of dice to roll to get the specific number they need. Here is the perfect opportunity to introduce and illustrate the simplest concepts of probability.

The Triples Game

To begin

Each player or team is given a set of four dice and is assigned a three-by-three array of squares to complete with numbers. Each player or team rolls a die. The player rolling the highest number plays first.

The play

1. The first player chooses to roll one, two, three, or four dice and rolls them. The player determines the sum shown on the thrown dice.
2. The player completes the turn by writing the sum of the dice in a position in the team's array. If one die is thrown, that number is recorded.
3. The play passes to the left. The next player chooses the number of dice to be rolled, rolls, then adds and places the sum in a position in his or her team's array.
4. No number may be shifted after it has been recorded.
5. Every sum determined by the play must be placed in the player's array.
6. The play continues until all squares in the assigned arrays are filled with numbers.

Scoring

Points are awarded as follows: 25 points for every addition fact in a row; 30 points for every subtraction fact in a column; and 50 points for every multiplication fact on a diagonal. If no facts are created, no score is given. Some examples of filled arrays and their scores are shown in figure 10.2.

9	6	14
1	9	2
8	3	11

25 - addition
30 - subtraction
_____
55 - total points

4	11	9
7	2	3
5	9	8

30 - subtraction
50 - multiplication
_____
80 - total points

12	6	13
3	4	7
5	2	6

25 - addition
30 - subtraction
30 - subtraction
_____
85 - total points

5	9	4
7	3	10
12	6	17

25 - addition
30 - subtraction
50 - multiplication
_____
105 - total points

Fig. 10.2

Bonuses

A bonus can be offered as a motivation for rolling higher numbers that are less likely to occur—10 points for sums higher than 16, 15 points for differences greater than 6, and 20 points for products higher than 14.

Variations

When the playing procedure has been mastered, teams or players may choose to play more than one assigned array. (A maximum of four is suggested.) This makes a longer, more interesting game, since there is a greater choice of positions for the rolled numbers and since more scoring occurs.

Sample of Triples Game Play

Assume the squares in each array are lettered (a) through (i) for reference purposes. Team A and Team B will alternate plays in their respective arrays.

(a)	(b)	(c)
(d)	(e)	(f)
(g)	(h)	(i)

First roll: Team A rolls two dice for a sum of 6 and places it in square (b). Team B rolls one die for a 3 and places it in square (e).

Team A
	6	

Team B
	3	

Second roll: Team A rolls one die for a 2 and places it in square (e). Team B rolls two dice for a sum of 8 and places it in square (a).

	6	
	2	

8		
	3	

Third roll: Team A rolls a die for a 1 and places it in square (g). Team B rolls four dice for a sum of 18 and places it in square (c).

	6	
	2	
1		

8		18
	3	

Fourth roll: Team A rolls two dice for a sum of 5 and places it in square (a). Team B rolls two dice for a 9 and places it in square (f).

5	6	
	2	
1		

8		18
	3	9

At this point in the game, Team A could complete an addition triple by rolling and placing an 11 in square (c), a subtraction triple with a 4 in square (d) or (h), or a multiplication triple with a 10 in square (i). Team B could complete an addition triple with a 10 in square (b) or a 6 in square (d), a subtraction triple with a 9 in square (i), or a multiplication triple with a 24 in square (i).

Fifth roll: Team A rolls a 10 with three dice and places it in square (i). Team B rolls a 9 with two dice and places it in square (i).

5	6	
	2	
1		10

8		18
	3	9
		9

Sixth roll: Team A rolls a 7 with two dice and places it in square (f). Team B rolls a 10 with two dice and places it in square (b).

5	6	
	2	7
1		10

8	10	18
	3	9
		9

Seventh roll: Team A rolls an 8 with two dice and locates it in square (c). Team B rolls a 7 with two dice and places it in square (h).

5	6	8
	2	7
1		10

8	10	18
	3	9
	7	9

Eighth roll: Team A rolls a 4 with one die and places it in square (d). Team B also rolls a 4 with one die and places it in square (d).

5	6	8
4	2	7
1		10

8	10	18
4	3	9
	7	9

Ninth roll: Team A rolls two dice for a 7, which goes in square (h). Team B rolls one die for a 5, which goes in square (g).

5	6	8
4	2	7
1	7	10

8	10	18
4	3	9
5	7	9

The final completed arrays and their scoring are shown in figure 10.3.

5	6	8
4	2	7
1	7	10

Team A Scores
30 – subtraction
50 – multiplication
—————————
80 – total points

8	10	18
4	3	9
5	7	9

Team B Scores
25 – addition
30 – subtraction
30 – subtraction
—————————
85 – total points

Implementation

To introduce the game to a class, divide the class into two teams. Give one set of four dice to each team. Explain or read the rules to them. Illustrate a scoring triple of each kind as shown in figure 10.1. Suggest limits by asking the class to find the highest and lowest numbers that can be rolled using one, two, three, and four dice. Keeping a note of these limits will help slower students to play more skillfully. Begin play as a class with the teacher's direction. Two games of play will give each of thirty-six students a chance to roll, add, and position a number. After completing the class games and scoring the results to determine the winning team, individuals can play in pairs.

As students are faced with choosing the number of dice to be thrown when they know they need a specific number, debates will inevitably arise about the best means of play. Introduce the probability sheet (see fig. 10.4) at this point. This sheet provides a framework for computing the probability of rolling a desired number with the different numbers of dice permitted. When completed (see fig. 10.5), the sheet provides a rational basis for selecting the number of dice to use in play. Although the completed sheet can be given to the students, most classes will benefit from computing the columns for one die and two dice by themselves or with the teacher's guidance. Classes of high ability will appreciate the challenge of computing the column for three dice. The column for four dice is difficult, and answers should be supplied only as needed. The sum of each column is 1, a fact that provides a nice check. Proficiency in simplifying and renaming fractions is required in this activity.

Summary

The game play itself requires thinking ahead and gives considerable practice in the use of basic facts. It can be used with slower classes without the probability sheet. The probability sheet, when issued complete with answers, introduces the need to compare fractions to determine the best chance.

Triples is a simple game to teach and learn. It creates a real need for concrete information on probability. Other applications of mathematical skills occur as an added benefit.

Probability Sheet

Desired Number	Number of Dice Thrown				Total
	1	2	3	4	
1					
2					
3					
4					
5					
6					
7					
8					
9					
10					
11					
12					
13					
14					
15					
16					
17					
18					
19					
20					
21					
22					
23					
24					
Total	1	1	1	1	

Fig. 10.4. This chart, when completed, will indicate your chances of getting the desired number in the first column by using the number of dice shown at the top of each column. The sum of each column will be 1.

Completed Probability Sheet

Desired Number	Number of Dice Thrown				Total
	1	2	3	4	
1	$\frac{1}{6}$	0	0	0	
2	$\frac{1}{6}$	$\frac{1}{36}$	0	0	
3	$\frac{1}{6}$	$\frac{2}{36}$	$\frac{1}{216}$	0	
4	$\frac{1}{6}$	$\frac{3}{36}$	$\frac{3}{216}$	$\frac{1}{1296}$	
5	$\frac{1}{6}$	$\frac{4}{36}$	$\frac{6}{216}$	$\frac{4}{1296}$	
6	$\frac{1}{6}$	$\frac{5}{36}$	$\frac{10}{216}$	$\frac{10}{1296}$	
7	0	$\frac{6}{36}$	$\frac{15}{216}$	$\frac{20}{1296}$	
8	0	$\frac{5}{36}$	$\frac{21}{216}$	$\frac{35}{1296}$	
9	0	$\frac{4}{36}$	$\frac{25}{216}$	$\frac{56}{1296}$	
10	0	$\frac{3}{36}$	$\frac{27}{216}$	$\frac{80}{1296}$	
11	0	$\frac{2}{36}$	$\frac{27}{216}$	$\frac{104}{1296}$	
12	0	$\frac{1}{36}$	$\frac{25}{216}$	$\frac{125}{1296}$	
13	0	0	$\frac{21}{216}$	$\frac{140}{1296}$	
14	0	0	$\frac{15}{216}$	$\frac{146}{1296}$	
15	0	0	$\frac{10}{216}$	$\frac{140}{1296}$	
16	0	0	$\frac{6}{216}$	$\frac{125}{1296}$	
17	0	0	$\frac{3}{216}$	$\frac{104}{1296}$	
18	0	0	$\frac{1}{216}$	$\frac{80}{1296}$	
19	0	0	0	$\frac{56}{1296}$	
20	0	0	0	$\frac{35}{1296}$	
21	0	0	0	$\frac{20}{1296}$	
22	0	0	0	$\frac{10}{1296}$	
23	0	0	0	$\frac{4}{1296}$	
24	0	0	0	$\frac{1}{1296}$	
Total	1	1	1	1	

Fig. 10.5. This chart indicates your chances of getting the desired number in the first column by using the number of dice shown at the top of each column. The sum of each column is 1.

11. Random Digits and Simulation

Edward Silver

J. Philip Smith

CAREFULLY constructed laboratory experiences with probability concepts are not only desirable from a pedagogical perspective but also essential if students are to overcome the erroneous notions present in the thinking of most probabilistically naive persons. Such experiences are particularly appropriate at the elementary or early secondary level, before students undergo any formal study of probability. Here we discuss two types of activities aimed at the beginning student. The two activities serve as examples of the kinds of motivational classroom experiences that can help students overcome common misconceptions and gain an understanding of basic probabilistic and statistical concepts.

Spinning Random Digits

Two of the most fundamental notions of probability are those of "equally likely" outcomes and "independent" events. Yet, despite the basic importance of the concept of "equally likely," an examination of junior high school and secondary school textbooks reveals that almost all of them seem to assume a correct understanding of the term on the part of the student. Research suggests, however, that a great number of students do *not* possess an adequate knowledge of the term.

One classroom activity that gives students direct experience with equally likely outcomes, as well as with the concept of independent outcomes, is one that uses 0-1 spinners. After students have become acquainted with the 0-1 spinner shown in figure 11.1, they are presented with, say, four different sequences of 0's and 1's. An example of four such data sets is presented in figure 11.2. The problem is to guess which of the four sequences was actually generated by the 0-1 spinner shown in figure 11.1. One sequence was generated by this 0-1 spinner; the other three were produced by one or more "biased" spinners. Can you guess which sequence resulted from this 0-1 spinner?

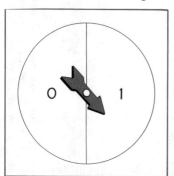

Fig. 11.1. A 0-1 spinner

70

Sequence 1: 0111110110111100100001011000110101001101110100001111001
Sequence 2: 110101010101011110111111110101011101101111010111101111
Sequence 3: 101101101010101110010010101010101011010110101001010101101
Sequence 4: 101111011100000110000000011100000011100110000111111101

Fig. 11.2. Four 0-1 sequences

Many students choose sequence 3, thus revealing a regrettable tendency to apply what has been called "the law of small numbers" [138]. Such students often reason, erroneously, that if 0's and 1's are equally likely to occur, then once a 0 has occurred, we should expect to see a 1 next; and once a 1 has occurred, then a 0 is likely to follow. The trouble with sequence 3 is that it is *too* rigorous about 0's following 1's and 1's following 0's. Over a small number of trials we should not expect such ideal behavior. A truly random sequence usually has far more runs—that is, strings of consecutive 0's or consecutive 1's—than sequence 3. In fact, sequence 3 was produced by the following rule: If a 0 occurs, put a 1 next with a probability of 3/4; if a 1 occurs, put a 0 next with a probability of 3/4.

To apply the rule for sequence 3 by using 0-1 spinners, start with spinner (a) of figure 11.3. If a 0 is obtained, use spinner (a) again. When a 1 appears, switch to spinner (b) and keep using it until a 0 appears; then return to spinner (a), and so on.

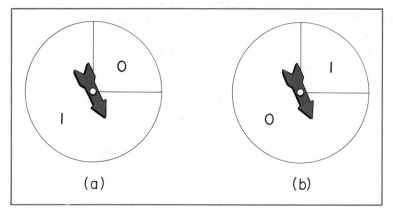

Fig. 11.3. Two 0-1 spinners

Incidentally, sequence 4 was produced by using spinners (a) and (b) but by switching only when the less likely digit occurred. Note that in sequence 4 the 0's are more often followed by more 0's and the 1's by more 1's. Sequence 4 has too many runs, whereas sequence 3 has too few. Sequence 2 was produced by using spinner (a) repeatedly—notice the dominance of 1's in the sequence. Sequence 1 was the sequence generated by the spinner shown in figure 11.1.

This type of exercise demonstrates the fundamental importance of runs in analyzing such data. Often, a mere frequency count is insufficient to force the data to reveal their secrets.

We shall now turn to a somewhat more sophisticated but highly motivational activity illustrating another use of the now-familiar 0-1 spinners.

Stimulating Simulation

Once students understand the behavior of 0-1 spinners, it is time to let them discover that such spinners behave much like many real-world situations. In fact, it is often easier to study the data from the spinners themselves rather than from the real-world situations they resemble. In other words, we can use the spinners to simulate more complex activity. Simulation is, in fact, a time-honored way to study phenomena that are too complex to analyze by, say, algebraic means.

Because our beginning students have studied little or no probability, the solutions to a great many problems are beyond their analytical abilities. Thus, simulation is an ideal method for them to use. As they gain acquaintance with a respected mathematical technique, they will be learning to think in probabilistic terms.

We shall discuss only one specific example, but it should be sufficient to suggest other such simulation activities appropriate for beginning students.

Consider the following situation: a section of a city has aging water mains and five pumping stations as shown in figure 11.4. Assume that at a particular fixed time each pumping station has a probability of failure of 1/2. At that time, what is the probability that water will flow from *A* to *B?*

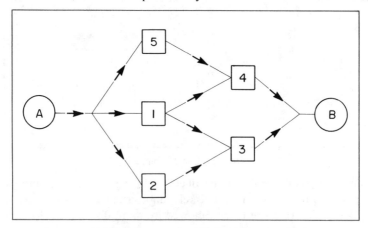

Fig. 11.4. Water mains and five pumping stations

The problem can be solved by algebraic methods and some knowledge of probability laws but will very likely be too difficult for beginning students.

Nevertheless, students can use simulation to obtain an estimate of the solution. In so doing, they will not only be gaining a grasp of the nature of random phenomena and familiarizing themselves with a powerful technique that can be used to analyze many real-life problems but will also be engaging in the kind of mathematical prediction and estimation that curriculum experts encourage us to teach.

How does the simulation proceed for our problem? Simply give five groups of students spinners of the type shown in figure 11.1. Each group, of course, represents a pumping station, and 0 outcomes will represent pumping failures. When each group has spun the spinner, the results are analyzed to see if a path exists for water to travel from A to B. After a reasonable number of trials, the class may want to give an estimate of the answer to the original question. (The answer is approximately 0.59.) If several groups work simultaneously on the problem, the number of independent trials mounts rapidly, and a large pool of data is collected.

This simulation clearly admits many variations—for example, vary the number of pumping stations or the probability of failure. The activity also raises many questions: How many trials is a "reasonable number" before estimating the solution? When is the answer "good enough"? What would happen if we removed pumping station 1 from the system? Do all the stations have to have the same probability of failure? These are all fine questions that you and your students could explore. If your students raise such questions, you are to be congratulated. We have seen many insightful questions emanate from similar exciting classroom simulations.

The sound, intuitive foundation of understanding fundamental concepts of probability, together with experience in simulations of real-world problems, should provide your students with an exciting springboard from which they can embark on the exploration of the fields of probability and statistics.

12. The Student Price Index

C. Gail Tibbo Lenoski

YOUNG people enjoy providing information about themselves and finding out facts about their peers. An interesting and multifaceted classroom project is to create a Student Price Index along the lines of the Consumer Price Index, a leading national indicator. The Student Price Index is designed to measure the increase over time in the prices of commodities that are particularly important to the student consumer.

The statistical concepts and methodologies that can be applied to this project are limitless. Students are exposed to data collection, questionnaires, analysis, and evaluation. There is a great requirement for systematization and organization of work and for executing a large number of simple statistical procedures and calculations. Teaching the quality and reliability of data are implicit to the project, as are indexing and time-series concepts. The exercise can include distribution, regression, correlation, chi square, and other topics; it also provides for a cohesive approach to a single project, with several stages to be carried out over an entire school term.

A logical concern to teachers is whether students can provide accurate information about their own expenditures and whether, in doing so, they have sufficient interest to provide valid data. Students should be able to recall any major expenditures they have made over the year. Purchases of a more routine nature should be easily recalled for a one-month period and can be multiplied by twelve to get an annual estimate. The Student Expenditure Survey includes a section on income for both the school term and the summer to help respondents recall jobs they have held, money they have saved, and their goals for certain purchases.

Parents could help their children recall the expenditures they have made and the income they have received. However, caution should be exercised to ensure that students report only purchases actually made by themselves, with money they have for discretionary spending.

Most, if not all, teenagers have a certain amount of disposable income, and the rise in prices and accompanying erosion of the purchasing power of the dollar is of concern to them. If a student's allowance is not increased in the face of rising prices or if the price of some elusive stereo equipment increases faster than the student generates income, the student is affected. A Student Price Index helps the class conceptualize in simple terms the movement of prices, the workings of the market, and something about the nation's economy!

From the methodology in the following sections, it may appear that the project is very time-consuming. It need not be so. A suggested time allotment is given in figure 12.1.

This is the basic amount of time required to conduct the project. A teacher who incorporates a wide range of statistical methods and techniques will find, of course, that the time requirement increases.

I should like to express special thanks to Jim Swift, Nanaimo District Senior Secondary School, Denis Desjardins, Assistant Director, Prices Division, Statistics Canada, and Dr. Miles Wisenthal, Administrator, Special Program for Chinese Scholars, Council of Ministers of Education (formerly Director General of Institutional and Public Finance Statistics, Statistics Canada) for their assistance, support, encouragement, and constructive criticism.

Activity	Number of 50-minute classes
Introduction to the project and discussion of the Student Expenditure Survey	One
Student Expenditure Survey	Homework to complete individual surveys, plus one class
Evaluation and examination of questionnaires	One
Basket selection table and weights table	Two
Standardization descriptions, pricing methodology	One, plus homework
Pricing	One
Indexing	One
Monthly pricing of items in stores and creation of monthly index	Homework, plus one class a month

Fig. 12.1. Time allotment for Student Price Index

One of the advantages of this activity as a long-term exercise is that it is not overly time-consuming. The students are given the opportunity of working on a project having an element of continuity, and they are encouraged to view statistics as a normal and integral part of life.

When the class has created a reasonable time series of price data (six months is ample) and calculated indexes, it is time to discuss what their results mean to them and how they can be interpreted.

The results may mean that some students will want to ask for an increase in their allowances. They may mean that some student has erroneously accused the cafeteria manager of raising prices every month. They may mean that it is becoming more expensive every month to ski. They may mean a great number of things, and students will, in all probability, be very enthusiastic about providing their own analyses of the price data and the ramifications of these data for themselves.

Methodology for Developing a Student Price Index

At the start of the school year, conduct a student expenditure survey (fig. 12.2) to collect information on where students spend their money and what commodities are most important to them in terms of expenditures. The survey, carried out by the students themselves, does not require names or addresses and thus protects the anonymity of the respondents. It is advisable to introduce the project and then give the students a month to

collect and compile information on their expenditures. More reliable and
more interesting price information will result.

Student Expenditure Survey

Age _____

Sex _____

Grade _____

School term

Do you have a paying job or jobs? _____ _____
                                     Yes   No

Average number of hours worked per week _____

Nature of work _____

Average income per week (from job)   _____

Other sources of income _____

Average income per week from other sources   _____

Total income during school term   _____
(Time period _____)

Summer

Last summer, did you work . . .

Full time   _____

Part time   _____

Occasionally _____

Average number of hours worked per week _____

Average weekly income from job(s)   _____

Other sources of income _____

Average weekly income from other sources   _____

Total summer income   _____
(Time period _____)

Total income (**school term** and **summer**)   _____

Fig. 12.2

Student Expenditures for the Past Year

(For items like food and other regular expenditures, estimate your expenditure over the past month and multiply this by 12 to obtain a figure for the whole year.* Try to recall all major expenditures and include in "others" those not covered in the listings.)

Please include only the items *you* paid for.

Total price should include sales tax where applicable.

Food Expenses

	(1) Frequency of Purchase	(2) Average Price	(1) × (2) Total Ex- penditure	Location of Purchase
Food at school*				
• Soup	_____	_____	_____	_____
• Sandwiches	_____	_____	_____	_____
• Hamburgers	_____	_____	_____	_____
• Milk	_____	_____	_____	_____
• Soft drinks	_____	_____	_____	_____
• Other (specify)	_____	_____	_____	_____
Between-meal food				
• Ice cream	_____	_____	_____	_____
• Potato chips	_____	_____	_____	_____
• Chocolate bars	_____	_____	_____	_____
• Soft drinks	_____	_____	_____	_____
• Gum	_____	_____	_____	_____
• Other (specify)	_____	_____	_____	_____
Restaurant meals**				
• Hot dogs	_____	_____	_____	_____
• Hamburgers	_____	_____	_____	_____
• French fries	_____	_____	_____	_____
• Pizza	_____	_____	_____	_____
• Chicken	_____	_____	_____	_____
• Other (specify)	_____	_____	_____	_____
Other (specify—no more than three items)	_____	_____	_____	_____
	_____	_____	_____	_____
	_____	_____	_____	_____

Total Food Expenditures   _____

*For the "food at school" section, monthly expenditure would be multiplied by 10, not 12, since there are only 10 months in the school year.

**Include here only if they constituted a meal. If not, they should be included with between-meal food.

Recreation Expenses

	(1) Frequency of Purchase	(2) Average Price	(1) × (2) Total Expenditure	Location of Purchase
Records				
• Singles	_____	_____	_____	_____
• Albums	_____	_____	_____	_____
• Tapes	_____	_____	_____	_____
Equipment				
• Radios	_____	_____	_____	_____
• Tape decks	_____	_____	_____	_____
• Record players	_____	_____	_____	_____
• Stereo sets	_____	_____	_____	_____
Attendance at—				
• Movies	_____	_____	_____	_____
• Discos	_____	_____	_____	_____
• Dances	_____	_____	_____	_____
• Spectator sports:				
• Hockey games	_____	_____	_____	_____
• Football games	_____	_____	_____	_____
Participation in sports events				
• Skiing	_____	_____	_____	_____
• Swimming	_____	_____	_____	_____
• Tennis	_____	_____	_____	_____
Recreation equipment				
• Skis	_____	_____	_____	_____
• Ski boots	_____	_____	_____	_____
• Tennis rackets	_____	_____	_____	_____
• Bicycles	_____	_____	_____	_____
• Skates	_____	_____	_____	_____
• Other (specify)	_____	_____	_____	_____
Other Recreation and Entertainment Expenses (specify)				
	_____	_____	_____	_____
	_____	_____	_____	_____
	_____	_____	_____	_____
	_____	_____	_____	_____
	_____	_____	_____	_____
Total Recreation Expenditures			_____	

Questions concerning the student's employment situation are useful to precipitate thinking along the lines of monies coming in and going out. They can also be used for later class analysis. A "nature of work" sheet generated by the class makes it easier to recall what they have earned. It would include babysitting, lawn mowing, waiting tables, and other typical student jobs. The "other income" section should include such monies as allowances and gifts of money. Data on income are not necessary for the actual construction of the price index, but they are useful for recalling expenditures as well as for putting the exercise in perspective for the students.

The expenditures part of the survey attempts to include items most commonly purchased by the student consumer. The class should be encouraged to suggest other items to be added to the list. Any frequently purchased item that has not been included in the questionnaire can be reported in the "other" category. Information on where the item was purchased is important for a later step, and students should be encouraged to complete this column as accurately as possible.

The class now examines the results of the expenditure portion of the survey. Spending patterns are likely to differ in some respects, and this should lead to lively class discussion. Quality and reliability of data should be introduced for discussion at this point. A "basket" of goods and services—not more than twenty—should be chosen from the expenditures section. These twenty items should represent those most important (in terms of proportion of income spent) and most frequently purchased by the class as a whole.

The class should generate from the completed questionnaires a table showing the number of students who purchase each item and the total amount spent on each item by all of them. The percent of students who buy each item should be calculated as well as *item expenditure as a percentage of total expenditures.*

Let us look at the "food at school" section (fig. 12.3) to determine how items in the basket should be selected. Assume there are thirty-five students in the class.

For simplicity, the last column has been calculated on the basis of *total expenditure on food at school* instead of on *total expenditures.* In the classroom, the *total expenditures* would first be calculated by adding all category totals, and then that figure would be used to calculate *item expenditure as a percentage of total expenditures.* Examples throughout follow this simplified pattern.

Create a table such as the one in figure 12.3 to include each item on the questionnaire. Look at each item and its importance in terms of both the percent of students who purchase it as well as the proportion of total expenditure it represents. The twenty items that are selected should represent commodities popular with students and significant in terms of total monies spent by the class.

Basket Selection Table

Food at School

(School year 19____–____)

Class 12G, 35 Students

Food item	No. of students who purchased the item during the year	Percentage of total no. of students	Class expenditure on each item during the year	Item expenditure as a percentage of total expenditures
Soup	4	11%	$ 194.40	3.0%
Sandwiches	23	66%	1614.60	24.8%
Hamburgers	31	89%	2343.60	36.0%
Milk	28	80%	1058.40	16.3%
Soft drinks	30	86%	1296.00	19.9%
	Total expenditure on "Food at School"		$6507.00	100.0%

Fig. 12.3

In the basket selection table, only four students, or 11 percent, purchase soup, and thus soup represents only 3 percent of total food expenditure. Soup, then, should probably be excluded from the basket we select. The students themselves will have to interpret the results from this table and decide which twenty items are the most appropriate to include. Consideration must be given to the ease with which a given item can be priced, its seasonality, and whether it is only a fad.

The next step is to determine the weight, or importance, of each of the items—that is, to define the proportion of total expenditures spent on each item in the basket. Return to the example of "food at school," this time eliminating soup. Create a weights table (fig. 12.4) for the basket. The

Weights Table

Food at School

Food item	Amount spent by students	Expenditure on each item as a percent of total expenditure	Weights
Sandwiches	$1614.60	26%	26
Hamburgers	2343.60	37%	37
Milk	1058.40	17%	17
Soft Drinks	1296.00	20%	20
Total expenditure for food at school	$6312.60		

Fig. 12.4

total expenditures figure then becomes the sum of expenditures on those twenty items. The percent of total expenditures spent on each item is the importance, or *weight,* of that item.

Pricing the items in the basket is the next step. For each item, a description should be made (fig. 12.5). This is to ensure that identical or equivalent items are priced each month. This is very important. If the price of a hamburger increases by ten cents but a slice of cheese has been added, there is not a true ten-cent increase, since the quality has also changed. In pricing a commodity such as a stereo record, students might include in the description "one of the top ten albums" to ensure that consistency is maintained in quality and value.

Hamburger

1/4 pound meat

bun, toasted

lettuce

tomato

onion

condiments (ketchup, mustard, pickles, relish)

Fig. 12.5

From the "location of purchase" section of the survey, the class should select the stores and facilities they normally patronize. A list of these stores is prepared, and each student is assigned responsibility for pricing one or two commodities, visiting the same stores each time they collect prices. To ensure standardization, a student should price the same items every month. Thus, if Mike, for instance, is assigned to collect prices in the record store, he is responsible for ensuring that he prices a record of the same quality each month. Optimally, several price quotations should be obtained on each item (that is, hamburgers should be priced in several different popular stores).

Pricing should take place once a month at the same time—for example, on the first Monday and Tuesday of each month. Each month's prices for an item for all outlets should be used to calculate a single average price. There will, however, be cases, particularly in small centers, where only one outlet will carry a particular item (for example, stereo records). In this event only the one price quotation will be used.

Students should actually visit stores to collect prices and should include sales tax where it applies. When an item is on sale, or presented as an advertised special, the reduced price is used, providing there is sufficient quantity of the sale item to make it widely available. A pricing chart (fig. 12.6) is helpful in organizing price collection.

Pricing Chart						
		Price (tax included)				
Item	Location	Month 1	Month 2	Month 3	...	Month 12
Stereo record albums	The Music Man	7.72	7.97	7.69		8.50
	Fishers	7.49	7.69	7.99		8.79
	Tune Town	7.29	7.59	7.99		8.49
	Average price	7.50	7.75	7.89		8.59
8-track tapes	Tune Town	9.45	9.45	9.59		10.25

Fig. 12.6

Now the teacher has an opportunity to introduce the concept of indexing and to explain the rationale behind using indexing in this project. If a stereo record costs $7.50 the first month pricing takes place—the base month—the base value of 100 is assigned to that price. Calculations of indexes for subsequent months are made along the lines of the chart in figure 12.7. After five months, the price of stereo records is 9.2 percent higher than at the beginning of the period. Indexes are calculated for each of the other commodities in a similar fashion.

Stereo Records			
Month	Average price	Percent increase in price from month 1	Index
1	7.50		100.0
2	7.75	3.3%	103.3
3	7.89	5.2%	105.2
4	7.99	6.5%	106.5
5	8.19	9.2%	109.2

Fig. 12.7

The twenty individual price indexes that result can then be averaged into an all-items Student Price Index.

Suppose after pricing for five months, the indexes in figure 12.8 result.

Thus the students have created an index that represents the change in the cost of selected student-purchased items over a specific time period.

Items	Index month 5 Base = 100	Weight	Weights multiplied by indexes
Sandwiches	109.1	26	2 836.6
Hamburgers	118.3	37	4 377.1
Milk	103.5	17	1 759.5
Soft drinks	107.4	20	2 148.0
			11 121.2

All-items index = 11 121.2 divided by 100 (reference period = 100)
               = 111.2

Fig. 12.8

Summary

This project has been presented to, and discussed with, secondary school students with very positive results. Suddenly numbers became exciting as the students became involved in an activity that was real, personal, and meaningful. Instead of sitting at a desk, they were learning while visiting stores, reflecting on their own finances, and discussing the project with others. A subtle form of learning experience took place; the students understood the front-page newspaper articles on inflation, and they had been exposed to economics as well!

It is the responsibility of educators to make learning as exciting and interesting as possible. Through projects such as this one, that aim may be achieved.

13. The Statistics Odds Room

Thomas E. Obremski

CLASSICAL motivating examples in the study of probability involve rolling a die repeatedly, tossing a coin repeatedly, choosing one or more balls from cans containing different proportions of colored balls, and other similar devices. Self-motivated students can learn much about probability

83

simply by experimenting with some of these devices. The classroom teacher can interest less able students in the study of probability by incorporating some of these classical devices into goal-oriented situations or games.

The incorporation of sound probabilistic models into games is the main objective of the Statistics Odds Room—a collection of games of chance operated, carnival style, as part of Ohio State University's annual High School Day. On entering the room, participants are given a fixed number of chips and invited to try their luck. The games are generally designed to reward good intuition so that the odds of winning are slightly in favor of the perceptive student.

Several of these games have been very well received and have been kept in the routine from year to year. A careful description of four of these games is included in this article. I believe they can be applied in any mathematics classroom at the sophomore level or higher. Included in each description are remarks concerning concepts involved in the play of the game, in figuring optimal strategies, and in varying the original game to obtain new games using the same basic materials.

One of the secondary objectives in presenting a series of already-developed games is to encourage teachers and students to formulate their own games and to try them out in class. Some of the general criteria used in our game construction are given at the end of this essay. They are offered simply as guidelines for the prospective game maker.

Description of the Games

Random Walk

Number of players: Any number, limited only by access to the layout

Equipment: Ten coins (dimes, for example) and the layout as shown in figure 13.1

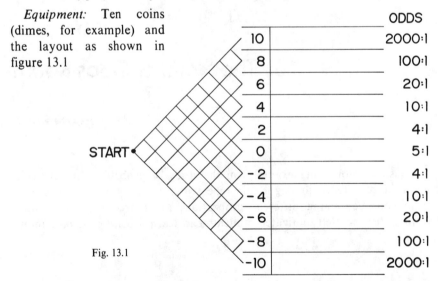

Fig. 13.1

Object: To choose the finishing point of the path of the ten coins

The play:

1. Each player, playing against the house, bets on one or more finishing points $(-10, -8, \ldots, 0, +2, \ldots, +10)$.

2. The house then tosses the ten coins in sequence. If the first toss is a head, the house places a coin on the next spot up and forward from START; if a tail is obtained, the next spot down and forward. Similarly, the second coin is placed forward of the first coin and on either the next spot up or down, depending on a toss of heads or tails. The process is repeated until the ten coins are used up, thereby creating a path of coins. If, for example, the tosses result in the sequence H H H T H T T H T H, the board would look like the one in figure 13.2. The person who has bet one chip on number 2 would, at 4:1 odds, receive four chips in addition to the one that was bet. All other players lose.

Fig. 13.2

Concepts involved:

1. *Independence of trials.* The outcomes of the first six trials give no information about the possible outcome of the seventh trial.

2. *Binomial probability function.* The total number of heads in the ten tosses varies from zero to ten. The probabilities with which these possible totals will occur are given by the binomial probability function. Each of the probabilities has the denominater 2^{10} and a numerator that is one of the coefficients of the expansion of $(x + y)^{10}$.

3. *Limitation of range.* The endpoint can be only an even number between –10 and 10.

Optimal strategy: With the odds set as in figure 13.1, the long-run optimal strategy is to bet on the 10 and the –10, since the expected payoff is larger than the bet. From a practical point of view, however, this strategy is optimal only if one intends to play an extremely long time. A better, short-run strategy is to bet on zero, which will occur almost one-fourth of the time and yet pays five to one, a very handsome payoff indeed.

Variations of the game: Side bets could be made concerning the first "peak" (H followed by T) in the path or the longest run of heads occurring in the tossing of the coins. Also of interest is whether or not the entire path will lie above the horizontal line through the starting point; side bets could be taken here at about 10:1 odds. Note that if the number of coins is changed to an odd number, only odd endpoints are possible.

Hypergeometric Horseracing

Number of players: 4

Equipment: Four large cans; 32 balls—16 black (B) and 16 white (W); one racetrack with four colored markers for horses. The track contains seven positions up to the finish line and three beyond it.

Object: To advance your horse beyond the finish line to a position farthest from START

The play: The entry fee is four chips from each player. In each round of play, each player in turn chooses one of the cans shown in figure 13.3, draws the number of balls specified there, and advances her or his marker the specified number of spaces. The balls are replaced for the next player. All players have the same number of turns and the same choice of cans. The winner is the person farthest from START at the end of the round in which at least one marker is past the finish line. The winner receives 12 chips; players who are tied divide the chips equally.

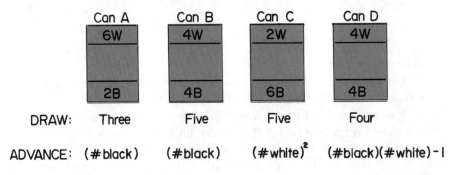

	Can A	Can B	Can C	Can D
	6W	4W	2W	4W
	2B	4B	6B	4B
DRAW:	Three	Five	Five	Four
ADVANCE:	(#black)	(#black)	(#white)^2	(#black)(#white) − 1

Fig. 13.3

Concept involved: Sampling without replacement. After choosing a can, the student is then interested in the number of black and white balls he or she will draw. The distribution of colors of the balls is governed by the hypergeometric probability distribution. After some practice, the students will notice that some cans are high-risk and some allow one to plod on to the finish line. If the sampling were done with replacement, the governing distribution would then be binomial.

Optimal strategies: Clearly can C is better than can A. Can B has the largest expected advance and probably should be used to begin with. However, if one needs a "4," the best chance is can C, which has about a 36 percent chance. If one needs a "3," the best bet is can D, which yields a "3" more than 50 percent of the time. Beware, however, that drawing from can D can result in a backward move. The skilled player should then view cans B, C, and D as means of achieving different goals, depending on the game situation at that player's turn.

Variations of the game: The proportions shown in figure 13.3 could be changed in some or all of the cans; the strategies would then change. The length of the racetrack should be altered to allow about three rounds for each race.

Scarlet and Gray

Number of players: Any number, limited only by access to the layout

Equipment: One die, a marker for each player, and the layout as in figure 13.4

Row Labels		BET	
Option 1	Option 2	(Scarlet) or	(Gray)
3, 4	ODD	6	2
1,2,5,6	EVEN	0	4

Fig. 13.4

Object: To obtain the largest payoff as indicated in the payoff matrix in figure 13.4

The play: Players compete with the banker. The option, 1 or 2, is specified at the beginning by the banker. The entry fee for each player is three chips. Each player selects a column by placing his or her marker on either "Scarlet" or "Gray." The banker rolls the die. Under game option 1, if the die comes up 3 or 4, then the payoff is either 6 or 2, depending on whether the player's marker is on scarlet or gray. Similarly, the payoff is either 0 or 4 for a roll of 1, 2, 5, or 6. Under game option 2, the row is selected according to whether the number on the rolled die is odd or even.

Concepts involved:

1. *Fair game.* Option 2 yields a fair game, one in which the player's expected return equals the bet. Option 1 is not a fair game; the player may have the advantage by selecting gray.

2. *Risk.* Under option 2, both scarlet and gray have expected payoffs of three chips. The choices differ drastically, however, in that the actual payoffs for scarlet differ a great deal; hence, the choice of scarlet involves more risk. The connection between risk and variability of the payoffs should be stressed when discussing the game.

Optimal strategy: Under option 1, bet gray. Under option 2, there is no optimal strategy.

Variation of the game: The four payoffs in figure 13.4 as well as the entrance fee can be varied to illustrate both fair and unfair games.

Bayesball

Number of players: 4

Equipment: Fifteen cards—ten labeled GO, three labeled STOP, and two labeled RETURN; a layout for a baseball diamond including at least the four bases; one marker for each player

Object: To advance one's marker around all the bases

The play: The entry fee is four chips. Players compete against each other, taking turns. All players have the same number of turns. Each player begins each turn with a complete, well-shuffled "deck" (see *Equipment*). At the start of a turn, each player may turn up zero to five cards so that everyone can see them. The player then selects a card from the remainder of the deck. If the card selected is STOP, the player's turn is over; if RETURN, the player returns to home plate and the turn is over; if GO, the player advances his or her marker to the next base and has the option of choosing another card from the remainder of the deck or passing the turn to the next player. The winner receives twelve chips; players who are tied divide the chips equally.

Concepts involved:

1. *Conditional probabilities.* The probability of obtaining a GO on a particular choice, given that four GO cards are showing and nine cards remain in the deck, is six out of nine, or 2/3. Conditional probabilities are of extreme importance in an area of statistics called Bayesian statistics after Thomas Bayes (1702–1761).

2. *Inference*. The players should choose to turn up all five cards allowed in order to try to derive some information about the unplayed cards. Deducing information about an unknown population from a sample is the basis for statistical inference.

Optimal strategy: A "perfect" hand to start with would have all three STOP cards and both RETURN cards turned up. The player could then proceed around the bases with no risk. This event will occur with the probability of 1 in 3003 (use the hypergeometric formula to derive this). Other strategy depends on the strength of the opposing players and on the player's relative position at any given point in the game. Generally it is best to gamble early or when behind, or both. The player going last has a definite advantage.

Variations of the game: Defining "winner" as the *first* one around the bases removes the advantage of going last. In this variation ties are not possible. The order in which the players proceed can be determined in advance by some randomizing device—highest roll of a die, for example. A two-person game with actual innings similar to real baseball can be played, regarding the "stops" as outs. This is not a particularly good group project, however.

Guidelines for the Prospective Game Maker

Some criteria used in the construction of the games are listed below. From an educational point of view the game should—

- incorporate one or more of the classical probability experiments, such as rolling a die or choosing balls from cans;
- contain some chance events whose probabilities could be calculated by an interested student after some instruction in the basic probability models and the rules of probability;
- point up some strategies that are apparently better than others but need not have an overall optimal strategy;
- be, if at all possible, a simplified version of the type of decision making that might be encountered in more practical situations, so that some transfer of learning can take place.

From an administrative or logistic point of view—

- a game should take three minutes or less to play;
- the rules should be relatively simple;
- the game should accommodate more than one player at a time;
- the materials should be readily available or, better yet, constructible.

14. Misconceptions of Probability: From Systematic Errors to Systematic Experiments and Decisions

J. Michael Shaughnessy

> This branch of mathematics [probability] is the only one, I believe, in which good writers frequently get results entirely erroneous.
>
> Charles Sanders Peirce

ANYONE who has had the opportunity to introduce students to concepts in elementary probability (or statistics) is probably aware of some of the misconceptions these students possess. For example, students often believe that the probability of getting three tails in four coin tosses is 1/4. The outcomes seem to be 3H, 2H and 1T, 1H and 2T, and 3T. The misconception arises from the belief that all these outcomes are equally likely to occur. Students may also feel that the probability of a tail is greater on the seventh toss after a run of six heads. This misconception, known as the *gambler's fallacy,* occurs because people expect that *even short runs* of the coin toss experiment should reflect the theoretical 50:50 ratio of heads to tails.

Other misconceptions arise because people are naive about how fast the partial products can grow or decay in counting problems. For example, people tend to greatly underestimate the number of possible batting orders for a nine-member baseball team (362 880). Also, when asked about the "birthday problem," most of us greatly overestimate the number of people needed to make the probability 1/2 that at least two of the people have the same birthday. In fact, many people will respond, "You must have at least 180 people to ensure a probability of 1/2 for matching birthdays."

Some of our misconceptions of probability may occur just because we haven't studied much probability. However, there is considerable recent evidence to suggest that some misconceptions of probability are of a psychological sort. Mere exposure to the theoretical laws of probability may not be sufficient to overcome misconceptions of probability. Cohen and Hansel [20], Edwards [29], and Kahneman and Tversky [63–66] are among those psychologists who have investigated the understanding of probability from a psychological point of view. The work of Daniel Kahneman and Amos Tversky is especially fascinating, for they attempt to categorize certain types of misconceptions of probability which they believe are systematic and even predictable. Kahneman and Tversky claim that people estimate complicated probabilities by relying on certain simplifying techniques. Two of the techniques they have identified are called *representativeness* and *availability*. We shall explore these two techniques in more detail and discuss some implications for teaching probability and statistics in the schools.

Making Estimates for Probabilistic Events through Representativeness and Availability

Representativeness

Those who estimate the likelihood of an event on the basis of how similar the event is to the population from which it is drawn or on how similar the event is to the process by which the outcomes are generated are using *representativeness*. For example, a long string of heads does not appear to be representative of the random process of flipping a coin, and so those who are employing representativeness would expect tails to be more likely on subsequent tosses until things "evened out." Below are several questions designed to test a subject's reliance on representativeness. The results are from a random sample of eighty college undergraduates prior to a course in probability. The numbers represent the frequency of responses.

R_1: The probability of a baby's being a boy is about 1/2. Which of the following sequences is more likely to occur in having six children?

   a)BGGBGB   b)BBBBGB   c) About the same chance for each
                                       of these sequences

Give a reason for your answer.

Responses to R_1		
BGGBGB	BBBBGB	Same
50	2	18

An overwhelming number of students chose the sequence BGGBGB to be more likely to occur. Their reasons often indicated that they felt this sequence fit more closely with the 50:50 expected ratio of boys to girls. These students are using representativeness. BGGBGB appears more representative of the population ratio than BBBBGB. Of course, both sequences are equally like to occur, with probability

$$(1/2)^6 = 1/64.$$

R_2: What is the probability that among six children three will be girls? (Same assumptions as R_1.) Give a reason for your answer.

Responses to R_2			
(Probability)	1/2	20/64	Other
(Frequency)	46	1	18

This was an open-ended question. The students could estimate any probability for the outcome. However, an extraordinary number of them said the probability was 1/2. It appears that they considered the probability of having a girl on one trial (1/2) to be a "representative" estimate for the ratio of girls in multiple trials. In fact, the chance of having three girls is 20/64.

R_3: The chance that a baby will be a boy is about 1/2. Over the course of an entire year, would there be more days when at least 60 percent of the babies born were boys—

   a) in a large hospital   b) in a small hospital   c) makes no difference

Give a reason for your answer.

Responses to R_3		
Small	Large	No Difference
17	15	48

Many of the students felt that the distribution of boys in a sample of babies should be the same regardless of the hospital size. Indeed, the effect that sample size has on the percentage of boys born was not a consideration in many of the responses. However, there is a good chance of at least 60 percent boys on a given day if the hospital has only, say, three babies a day! Students tended to use the 50:50 distribution of boys to girls in the population as a representative estimate, regardless of sample size. The result was an abundance of "no difference" responses to this question.

The use of representativeness by students who have no prior knowledge of probability is prevalent in the items above. Representativeness can be shown to account for errors in prediction that arise from the following attitudes:

   1. Insensitivity to prior probabilities and disregard for population proportions

2. Insensitivity to the effects of sample size on predictive accuracy
3. Unwarranted confidence in a prediction that is based on invalid input data
4. Misconceptions of chance, such as the gambler's fallacy
5. Misconceptions about the tendency for data to regress to the mean

Availability

When people tend to make predictions about the likelihood of an event based on the ease with which instances of that event can be constructed or called to mind, they are relying on *availability*. For example, if a person is asked to estimate the local divorce rate or to estimate the probability of being involved in an automobile accident, the frequency of his or her personal contact with these events (perhaps through friends or relatives) may influence the probability estimates. One who has recently been involved in an accident might estimate a high probability for the accident rate. One who has many friends and acquaintances who are divorced has more instances of divorce "available" and may estimate a higher local divorce rate than someone who does not know many divorced people.

Availability causes systematic bias in probability estimates because people tend to believe that those outcomes that can easily be brought to mind will be more likely to occur.

Below are the results of several items that were used to assess students' reliance on availability prior to a course in probability.

A_1: Consider the grids below.

Grid A	Grid B
XXXXXXXX	XX
XXXXXXXX	XX
XXXXXXXX	XX
	XX
	XX
	XX
	XX
	XX
	XX

Are there

a) more paths possible in grid A?
b) more paths possible in grid B?
c) about the same number of possible paths in each grid?

Give a reason for your answer.

(A path was carefully defined as a polygonal chain of line segments, starting at the top row and proceeding to the bottom row and meeting one and only one symbol in each row. Examples were provided.)

Responses to A_1		
Same	Grid A	Grid B
15	53	8

There are, in fact, the same number of paths in each grid, since $8^3 = 2^9 = 512$. However, a majority of the students felt that more paths were possible in grid A. The reasons students gave for choosing grid A over grid B included, "There are more Xs in grid A" and "It is easier to draw a path in grid A." Paths just seem easier to construct in grid A, that is, there appear to be more paths *available* in grid A.

A_2: A person must select committees from a group of 10 people. (A person may serve on more than one committee.) Would there be

a) more distinct possible committees of 8 people?

b) more distinct possible committees of 2 people?

c) about the same number of committees of 8 as committees of 2?

Give a reason for your answer.

Responses to A_2		
Same	Committees of 2	Committees of 8
19	47	7

There was a strong tendency for the students to choose "committees of 2." They used availability because examples of committees of two are easier to construct or call to mind than committees of eight. In fact there are the same number of committees of two as committees of eight. Picking a committee of two is the same as picking a noncommittee (those left out) of eight.

These results support the contention that people who do not know much about probability and statistics (i.e., are *combinatorially naive*) will tend to rely on such guessing principles as representativeness and availability to estimate the likelihood of events. However, *even people who know* something about probability and statistics may rely on these principles rather than on their knowledge of probability. For example, social scientists with a substantial background in probability and statistics were subject to the same types of misconceptions as people who were naive about probability. Furthermore, I have found evidence that mere exposure to probabilistic concepts does not prevent students from relying on representativeness and availability [112]. They may have the same types of misconceptions after studying probability as they did beforehand.

Developing Probabilistic Intuition

A case for teaching probability and statistics in the schools, perhaps as early as possible, has been eloquently set forth by several authors. These authors have pointed out the need for people to learn about probability and statistics just to be able to make reasonable decisions (as consumers or voters or even in choosing a career) on the basis of the mounds of data and probabilistic statements that confront them. It is indeed difficult to be a well-informed citizen or a critical consumer without a good understanding of probabilistic statements.

These misconceptions of probability provide a second crucial reason for the early introduction of probabilistic thinking into our school mathematics curriculum. The evidence suggests that when students enter college, most of them are *(a)* unfamiliar with probability and *(b)* subject to misconceptions of probability that are deeply entrenched and therefore hard to overcome. Furthermore, the type of introduction to probability that these students are likely to encounter in a college course is almost certain to be formalistic. In college probability concepts are usually introduced through the language of sets and sample spaces in an abstract model of statistical probability. I have gathered evidence that suggests that an initial formalistic approach to probability is unlikely to help students overcome misconceptions. However, if probability is introduced first through experiments, students appear to have more success in overcoming their probabilistic prejudices. Thus a second reason for introducing probability in the schools is to develop probabilistic intuition in children and adolescents. Fischbein [40] claims that science education in Western cultures has emphasized only the *necessary*, the deterministic aspects of science and has neglected the study of the *possible*, of uncertainty. Thus our intuition of probabilistic thinking has been distorted by an overemphasis on deterministic models, such as natural selection, Newton's laws of motion, or a deductive study of geometry from axioms.

School mathematics programs have an excellent opportunity to help students develop their probabilistic intuition and to promote their growth as informed citizens and critical consumers.

Teaching Probability to Overcome Misconceptions

In order to develop adequately probabilistic intuition and statistical literacy in our students, I should like to suggest that we should make the study of probability and statistics an *integral, required* part of the school mathematics program and introduce probability to all students not later than junior high school.

Here are three recommendations for the study of probability and statistics in the schools:

1. Introductory probability and statistics should be activity-based, experimental probability.

2. Emphasis should be placed on simulation—both as a tool to model experiments and as a problem-solving technique.

3. We should sensitize our students to the misuses of statistics and encourage them to analyze probabilistic statements critically.

Let us elaborate on two of these recommendations.

Experimental Probability

There is evidence suggesting that students have a better chance of overcoming misconceptions of probability that are due to representativeness or availability if they are introduced to probability in an experimental, activity-based mode rather than in a formalistic lecture mode. Students can work in small groups, make guesses for the likelihood of events, perform an experiment, organize and analyze the data, and make estimates for the likelihood of events based on their data. More advanced students can perhaps attempt to build a theoretical probability model for the experiment and then compare theoretical predictions to experimental outcomes. The reasons for any discrepancies between experimental results and theoretical predictions, such as sources of bias in the data-gathering process, can be discussed.

Examples of probability activities that can be done by students working together in small groups follow. The first activity deals with equally likely outcomes, the second with outcomes that are not equally likely to occur.

Activity 1: Coin Flipping—Keep Your Head Up

Before doing this activity, write down your *best guess* for each of the following: If you flip 6 coins, what is the probability that you will get—
- *a)* 6 heads   *c)* 4 heads
- *b)* 5 heads   *d)* 3 heads

Perform this activity in groups of six students:

1. Flip 6 coins. Record the number of heads. Repeat about 50 times. Use your *data* to answer each of these questions.

   a) What is the probability of getting 6 heads? 5 heads? 4 heads? 3 heads? 2 heads? 1 head? 0 heads?

   b) What is the probability of getting at least 1 head? At least 2 heads?

2. Make a list of all possible outcomes for flipping 6 coins.

   a) Develop a mathematical model to find the *theoretical* probability for the outcomes of flipping 6 coins.

   b) What is the theoretical probability for getting at least 1 head? At least 2 heads?

  c) What are the assumptions of your mathematical model?

3. Compare the experimental probabilities from part 1 with the theoretical probabilities in part 2. How well do they agree? Make a graph to compare the experimental (observed) probabilities in 50 flips for 0 heads, 1, ..., 6 heads with the theoretical probabilities. Plot the two graphs on the same set of coordinate axes. Where is there close agreement between the two graphs? Where is there not close agreement? Why do you suppose this happens?

4. What assumptions have you made in your experiment when flipping the coins? What suggestions do you have to improve the experiment?

5. List any other comments, questions, observations, or reactions that you might have concerning this activity.

Activity 2: Thumbtacks—a Pointed Affair

Three different colors (red, gold, and silver in this example) of ordinary household thumbtacks are distributed, one tack to each person.

1. Do this part of the activity by yourself.

  a) First, write down your *best guess* for the probability that your tack will land upright when dropped.

  b) Devise some uniform way of dropping your tack, and drop it 75 times. Arrange your data in an array with *U* to indicate upright and *D* to indicate down.

  c) On the basis of the data you have collected, calculate the probability that the tack will land upright; that it will land down.

2. Do this part of the activity in small groups. Get a person in your group with one each of the three colors of tacks. Use the probabilities calculated in part 1 to list the probabilities that (1) the red tack will land up, (2) the silver tack will land up, and (3) the gold tack will land up. Drop the three tacks together and record the results. Before performing this experiment, write down your *best guess* for each of the following probabilities:

  a) All the tacks land up

  b) No tack lands up (what's another way to say this?)

  c) At least one tack lands up (what's another way to say this?)

  d) The red tack lands up

  e) Two tacks land up and one lands down

Drop the three tacks in some uniform way about 60 times. Record your data in an array by listing the results as triples; for example, *UDD* could stand for red tack up, gold tack down, and silver tack down.

Use the data you have gathered to calculate the experimental probabilities for the events in *a–e* in part 2. Compare these calculations to your guesses. Any surprises? What assumptions have you made in doing the experiment? What suggestions do you have to improve this experiment?

3. Do this part of the activity in small groups.

   a) Develop a mathematical model to assign theoretical probabilities to the outcomes of this experiment. First, list all possible outcomes for the experiment; then devise a way of assigning a probability to each outcome.

   b) Use your data from part 2 to determine experimental probabilities for each of the outcomes listed in 3*a*. Compare these experimental probabilities with the theoretical probabilities for the outcomes given by your model. Make a graph to compare your experimental probabilities with the theoretical probabilities. How well do the graphs agree?

   c) What assumptions have you made in your mathematical model? Does the model for the tack experiment differ in any way from the model for the coin experiment? If so, how? Is there any similarity between the two experiments?

   d) List any other comments, questions, observations, or reactions that you might have to this activity.

Activities such as these are aimed at correcting misconceptions. The tossing of six coins is equivalent to our representativeness problem with the sequences of six children. The tack activity provides experience with unequally likely outcomes so that students begin to realize that if there are N possible outcomes, this does not mean that each outcome has a $1/N$ chance of occurring. Furthermore, the activities are in two stages. The first stage deals with experimental probability. The second stage encourages students to *build their own* theoretical models for assigning probabilities. Activities in which just the first stage is done are more appropriate for the elementary and junior high school years, whereas the two stages work well with secondary students or in introductory college courses on probability.

Misuses of Statistics

Exposure to misuses of probability and statistics should help students confront their own misconceptions of probability; this would seem to be an essential ingredient in any introductory course on probability and statistics. Misuses of statistics can affect the human decision-making process, which in turn affects the course of human lives. Darrell Huff sums up the consequences of the misuses of probability and statistics in his book, *How to Lie with Statistics* [58]:

> So it is with much that you read and hear. Averages and relationships and trends and graphs are not always what they seem. There may be more in them than meets the eye, and there may be a good deal less.
>
> The secret language of statistics, so appealing in a fact-minded culture, is

employed to sensationalize, inflate, confuse, and oversimplify. Statistical methods and statistical terms are necessary in reporting the mass data of social and economic trends, business conditions, "opinion" polls, census. But without writers who use the words with honesty and understanding and readers who know what they mean, the result can only be semantic "nonsense." (p. 8)

Students can be asked to keep a log of misuses of statistics that they discover. Articles can be taken from newspapers or periodicals and analyzed for correct or incorrect use of statistics. Students in my classes have ferreted out misleading graphs, inflated percentages, biased samples, insufficient sample size, and verbose statistical quantitative descriptions that had little or no basis in fact.

Misuses of percent have been mentioned by many of my students. For example, a percent of cost increase or of profit can be conveniently inflated or deflated by merely changing the denominator units. One student reported that a recent tuition hike at her university was purported to be a 15 percent increase when in fact it was a 22 percent increase. The university had calculated the percent of increase by $I/(C + I)$, where I was the increase per credit hour and C was the cost *before* the tuition hike. The student calculated the increase by I/C.

Many examples of statistics that were based on a biased sample have been analyzed by my students. One found an article on page 1 of the largest-selling newspaper in a major city that mentioned that 95 percent of the people who were interviewed were against cross-district busing to achieve racial integration in the schools. A continuation of the article (on page 14) mentioned that 70 percent of the people interviewed *did not have children in the schools.* Another student criticized the charts published by bookstores and record stores to indicate the top ten books or records for the week. These figures are often based on sales to a particular class of people with special interests. Yet the list purports to reflect the population as a whole.

I have suggested that an approach to introductory probability that emphasizes experimental activities, simulation, and misuses of statistics can help to combat misconceptions of probability, such as those based on representativeness or availability, and can also help to develop probabilistic intuition. The experiments can be carried out by the students working in small groups while the teacher plays the role of organizer, diagnostician, devil's advocate, and critic. There are even several bonuses that a small-group, activity-based approach to probability can generate:

1. Students can see the variability of experimental results from group to group. This can lead to a discussion of sources of bias in the data-gathering process. It can also spark a future study of the concepts of variance and deviation.

2. The influence of sample size on experimental results can be investigated.

3. Peer teaching and interaction can be quite exciting in a small-group setting.

4. Misconceptions can be confronted by comparing guesses to experimental results.

5. Simulation can be introduced, providing a powerful tool to investigate expected value, decision making, and complex probability estimates from an experimental point of view.

6. Probabilistic intuition can be developed for a subsequent, more theoretical approach to probability.

7. It's a lot of fun for everyone, students and teacher, to study probability this way.

15. Some Statistical Paradoxes

H. E. Reinhardt

STATISTICAL methods," according to Cox and Hinkley [27], "are intended to aid the interpretation of data that are subject to appreciable haphazard variability." The methods are eclectic, and consequently it is often difficult to decide which of several ways of analyzing data is most appropriate. It is the purpose of this article to discuss instances in which different analyses lead to apparently paradoxical results and to resolve those paradoxes, at least partially. The paradoxes discussed here arise primarily from treating nonrandom samples as if they were random, or from misinterpreting probabilities that have been correctly computed.

This material has been used several times with eleventh-grade students after they have studied chapters 8 and 12 of Book 0 of the Comprehensive School Mathematics Program series [23]. The specific techniques used are (1) tree diagrams and Bayes' theorem for computing probabilities and conditional probabilities and (2) simulation for estimating probabilities and expected values. (Several other articles in this yearbook deal with these topics. "Paradoxes in Sampling," by Clifford Wagner, focuses on conditional probability, and articles by William Inhelder and by Ann Watkins and Kenneth Travers deal, respectively, with computer simulation and with Monte Carlo techniques.)

Bayes' theorem is the basic result showing how information changes probabilities. It can be conveniently discussed using tree diagrams. This method will be illustrated in solving the following problem:

A college requires prospective students to take an entrance examination as a basis for admission. The college estimates that 50 percent of the applicants are capable of completing a bachelor's degree. It estimates that 15 percent of the unqualified applicants pass the exam and 5 percent of those qualified fail it. All who pass the exam are admitted and none of the others are. What fraction of prospective students are admitted? If an applicant passes the exam, what is the probability that he or she is capable of completing a degree? Denoting by C the event "Applicant is capable of getting a degree," by A the event "Applicant passes the exam," and by C' and A' the respective complementary events, we summarize the information in the tree diagram of figure 15.1. Using the product rule for computing probabilities, we multiply the branch probabilities to obtain the path probabilities appearing in the right-hand column of the figure.

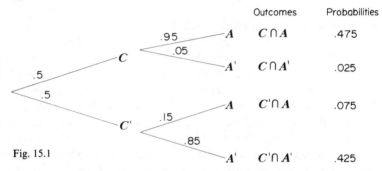

Fig. 15.1

The first and third paths are those representing admitted students, so that $P(A) = .475 + .075 = .55$. This is the answer to the first question; 55 percent of applicants are admitted. The first path shows that 47.5 percent of students are both capable and admitted, and so the conditional probability of C, given A, is $P(C|A) = P(C \cap A)/P(A) = .475/.55 = .86$. This is the answer to the second question; 86 percent of those admitted are capable of completing a degree. The computation could be made by using Bayes' formula rather than the tree diagram. Bayes' formula is, of course, implicit in the tree diagram. Without the exam, the probability is .5 that an applicant is capable of getting a degree. Knowledge that the applicant has passed the exam changes this probability to .86. In general, knowledge changes probabilities. Tree diagrams are a convenient way of determining the changes.

Simpson's Paradox

We describe this paradox, named for a statistician who gave a careful discussion of it, with a scenario involving a rookie, R, who is trying to break into a baseball lineup by replacing an old-timer, OT. He is told by the manager that the decision will be made on the basis of hitting ability, since both R and OT are equally proficient as fielders. The rookie feels assured that he will be the starting player because he is batting .224 and OT is batting only .186. (Both are "good field, poor hit" players.) However, the rookie is dismayed to learn that OT is designated to start the first game. He asks the manager why he is not the starter, since his average is better than that of OT.

"Well," explains the manager, "the opponents are using a right-handed pitcher today, and OT has a better average than you against right-handed pitchers. His average is .179 and yours is .162."

Satisfied but chagrined by the explanation, R waits until the opponent's pitcher is left-handed, for then, surely, he will get to start. But that day comes and OT is once again the starting player. The manager points out that OT is better than R against left-handed pitchers, too, batting .560 compared to R's .332.

The rookie is no longer satisfied with the explanation, and he asks how OT can be better than R against both left- and right-handed pitchers but poorer overall. The data shown in table 15.1 are persuasive. The arithmetic is correct, and careful analysis of the table may reveal the source of the paradox. The total batting percentage is determined by the percentages against the two types of pitchers but is determined differently for each of the two players. The overall percentage is a *weighted average* of the individual percentages, and the weights are determined by the number of times at bat against each of the types of pitchers. Although we expect the weighted average to preserve the order of the individual percentages, it does not.

TABLE 15.1
BATTING RECORDS OF TWO BASEBALL PLAYERS

	Times at Bat	Hits	Average
Totals	OT 4766	888	.186
	R 1276	286	.224
Against right-handed pitchers	OT 4675	837	.179
	R 809	131	.162
Against left-handed pitchers	OT 91	51	.560
	R 467	155	.332

Figure 15.2 shows tree diagrams for the old-timer and the rookie that formalize the preceding discussion.

The overall probability of a hit is given by the sum of the first and third path probabilities. The records against right- and left-handed pitchers appear in the tree diagrams as conditional probabilities in the second stage. To obtain path probabilities, these conditional probabilities must be

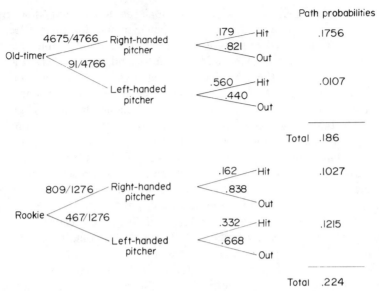

Fig. 15.2

multiplied by *different* probabilities from the first stage. If R and OT had batted against the two kinds of pitchers the same fraction of times, the paradox could not occur. (Bryan Wilson, in his paper "The First Shall Be Last" [142], uses a graphical representation that is informative when there is a natural order to the records to be combined.)

Although the story of the rookie is fictitious, the data are not. They are adapted from death rates for tuberculosis in 1910, as shown in table 15.2, given by Cohen and Nagel [21]. The death rates from tuberculosis for both whites and nonwhites were greater in New York than in Richmond. However, the overall death rate from tuberculosis was greater in Richmond than in New York. Public health officials might have reached very different conclusions, depending on which part of the table received their attention.

TABLE 15.2
DEATH RATES FROM TUBERCULOSIS, 1910

	Population		Deaths from Tuberculosis		Death Rate per 100 000	
	New York	Richmond	New York	Richmond	New York	Richmond
White	4 675 174	80 895	8 365	131	179	162
Nonwhite	91 709	46 733	513	155	560	332
Total population	4 766 883	127 628	8 878	286	186	224

Consider another situation where Simpson's paradox might arise. Suppose one wants to compare two school lunch menus for their effect on weight gain. A standard procedure is to use two random samples of children, giving one menu to one group and the second to the other. A statistician might find that the second menu gives greater average weight gain than the first, whereas the first gives greater weight gain for boys and girls separately. The vagaries of sampling include a higher fraction of girls in one sample than the other so that weighted averages reverse the implications of the results for individual sexes. Of course, that will not happen very often. Random samples are designed to give correct answers most of the time. However, if we anticipate that boys and girls will react differently to the diets, we are better off to force the samples to have the same representation of the sexes (by sampling boys and girls separately) rather than to depend on the fact that random samples usually give correct results.

The article by Bickel, Hammel, and O'Connell [6] contains another, legally perplexing instance of Simpson's paradox. Here is a simplified version, which does not obscure the essential difficulty: A university is accused of discriminating against female applicants for graduate school because it has admitted a higher proportion of male applicants than female applicants. However, in each of the individual colleges of the university, women and men are admitted in the same proportion. The paradoxical situation arises because most women applicants apply to the College of Arts and Sciences, which has the lowest acceptance rate, and the fewest women apply to engineering, which has the highest acceptance rate. Thus, at the college level where the decision is made, there is no evidence of discrimination, but affirmative action mandates the recruiting of women into engineering.

The Classification Paradox

In our preliminary example, we considered a typical classification scheme where university applicants are admitted or rejected. There are two possible decisions and consequently two types of errors—applicants who should be admitted are rejected, and applicants who should be rejected are admitted. If we think of the decision process as testing the statistical hypothesis that the applicant is capable of completing a degree (and is, hence, worthy of admission), the two errors are called, respectively, errors of the first and second kind. Clearly, a good classification procedure has small probabilities of the two kinds of error. We are inclined to think—and statistics textbooks strongly suggest—that the converse is true: if the two kinds of error are small, the classification scheme, or hypothesis-testing procedure, is a good one. That is the case with the preliminary example; most of the people admitted are capable of completing a degree and few of the capable ones are denied admission. This is only partially due to the small probabilities of the two kinds of error, as the following artificial example illustrates.

Suppose a diagnostic procedure (such as a skin test for tuberculosis) correctly diagnoses the presence of a disease in 99 percent of persons who have it and its absence in 99 percent of those who do not have it; the probability of each of the two kinds of error is .01. If the procedure is used in a clinic half of whose clients have the disease, then the test performs very well, as the tree diagram of figure 15.3 shows. (We write "+" for positive diagnosis, "−" for negative diagnosis, and "D" for presence of the disease.) The patients on the first and third paths—50 percent—are diagnosed as having the disease. Those on the first path actually have the disease, and so the probability that a person who is diagnosed as having the disease actually does have it is $.495/.500 = .99$.

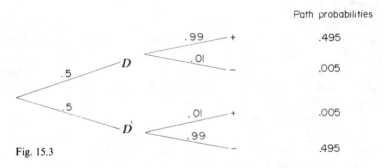

Fig. 15.3

In the clinic the procedure works very well. It could happen that public health officials, observing the clinical success, decide to administer the test to all schoolchildren. They may be surprised to find that the test no longer performs well. Suppose, for instance, that one child in 1000 has the disease. Then the appropriate diagram is the one in figure 15.4. Here the probability of positive diagnosis is $.000\ 99 + .009\ 99 = .010\ 98$. But of those so diagnosed, only a fraction, $\dfrac{.000\ 99}{.010\ 98} \approx .09$, have the disease. Paradoxically, the great majority of positives are *false positives!*

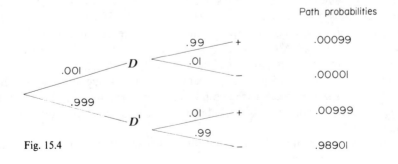

Fig. 15.4

There is no real paradox. The reason the procedure performs well in the clinic is not just because the two kinds of error are small but also because the procedure is used on a population with a large number of persons with the disease. In the population of schoolchildren, few have the disease, and so most of the positives are false. Galen and Gambino [43] have discussed a number of interesting actual medical situations in which most positive diagnoses are false positives.

The Inspector's Paradox

The validity of standard statistical methods requires a random sample. Many nonrandom samples are treated as random (often with disastrous results), and occasionally a sample thought to be random is not. In this section we analyze such a procedure. Similar paradoxes of sampling are discussed by Feller [39].

Arthur (A) and Barbara (B) are hired to inspect pennies at the mint. The procedure is to toss coins observing the "strings" of tosses ending in a head. For example, the sequence of tosses TTH TH TTH TTTH H consists of "strings" of length 3, 2, 3, 4, and 1. The tenth toss is contained in a string of length 4. For a fair penny, the theoretical average string length is 2. (The derivation of this fact is an appropriate exercise for students.)

A and B are told to choose a random string for each coin and use its length to decide if the penny is fair. Large discrepancies from the theoretical average of 2 are taken as evidence that the penny is biased. A decides to test the pennies after tossing them a while to "warm them up." In particular, he decides to look at the string of tosses that *includes* the tenth toss. B, inspecting the same coins, decides to use them after A has warmed them up. She will look at the string of tosses following A's string. The tests of the first coin give the sequence H H TH H H H TH TTTH TTH. The tenth toss starts a string, and so A's result is TTTH and B's is the final string, TTH. Table 15.3 gives the results of the first fifteen pennies; B's string is the last string and A's the next-to-the-last string and always includes the tenth toss. Even on the basis of such a small number of tosses, it is clear that A will conclude that the mint is producing coins biased toward tails and B will conclude that the coins are fair. The results for 1000 simulated unbiased coins give an average of 3.036 for A and 1.994 for B. It is a fact that A's procedure does *not* give a random sample and B's does. For instance, the only way A's string can have length 1 is for both the ninth and tenth tosses to be heads (so his probability is 1/4). B's string has length 1 if her first toss is a head (so her probability is 1/2). Roughly speaking, B's procedure is random because the coin has no memory, but since long strings have a greater opportunity to include the tenth toss than short ones, A's is not.

Arthur's friends suggest ways to improve his sample scheme. Claire sug-

TABLE 15.3
TESTING PENNIES

Sequence	A's Result	B's Result
H H TH H H H TH TTTH TTH	4	3
H TTH H H TTH TTH TH	3	2
TTH H TH H TTH TTTH	3	5
H H TTTTTTH TTTTTH H	6	1
H TTH TH H TH TTTH TTH	4	3
H TTH H H H H TTH TTTH	3	4
TH H TH H H H H H H	1	1
TTH H TH H H H TTTTTTTH TH	8	2
H H TH H H TH TH TTTH	2	4
TTH TH TTH H H H	1	1
TH TTH TH TTTH TH	4	2
H H TTTH H H TH H	2	1
H H H TTH H TTH H	3	1
H TTH TH H H TTTH H	4	1
H H H TTTH H H TH H	2	1

gests warming up the penny by tossing it until it comes up tails and start-ing a string with the next toss. Donald suggests tossing the penny for ten tosses and starting a string with the eleventh toss. Does either of these methods give a satisfactory sample? Or do both schemes suffer from the fact that Donald's may, and Claire's must, require counting to begin in the middle of a string?

Arthur Engel [32] has suggested the following amusing illustration of a related error in sampling procedure. Ask the students in the class to give the number of boys and of girls in their families. Tabulate the results separately for the boys and girls in the class and calculate the average number of boys per family and the average number of girls per family. Table 15.4 shows a typical result. The claim is that families tend to "run" to children of one sex, as the data superficially indicate. The difficulty is that the families of the boys in the class are not a random sample of fami-lies, nor are the families of the girls. *Query:* Are the families of the chil-dren in the class a random sample of families?

TABLE 15.4
FAMILIES OF BOYS AND GIRLS

	Average Number of Boys	Average Number of Girls
Families of boys	2.00	0.81
Families of girls	0.77	2.09
All families	1.22	1.22

Other Paradoxes

There are a number of other well-documented paradoxes. We mention them briefly and provide references to more complete discussion. Billstein

[7] considers the use of some of them at a more elementary level than that of this paper.

The nontransitivity paradox [8]

Our sense of choice suggests that preferences are transitive; that is, if A is preferred to B, and B to C, then A is preferred to C. But, paradoxically, that is not necessarily so. A nice example, using four dice, is a modification of one credited to Bradley Efron [45]. Die A has all 3s. Die B has four 2s and two 6s, die C has three 1s and three 5s, and die D has four 4s and two 0s. One can easily construct such dice and play the following instructive game. Let a student choose a die; then you choose one. The dice are rolled, and the person with the larger number showing is the winner. Since die A will beat B in two-thirds of all games, A is preferred to B. Similarly, B is preferred to C, and C to D, but D is preferred to A. Because of the non-transitivity, the person who chooses first is always at a disadvantage.

The clocking paradox [8]

We expect results from time trials to give us information about performances in races. For instance, two persons who do equally well in time trials will do equally well in a race, all other things being equal. However, consider A and B, who are mile racers. A runs the mile in 4.00, 4.01, 4.02, . . . , 4.09 minutes, each time having probability .1. In these races B runs in 0.01 minute less than A except when A runs a four-minute mile, and in those races B runs the race in 4.09 minutes. Thus B wins 90 percent of the races, but if the results are interpreted as time trials, A and B have equally good records. We could modify the example slightly to show that A can actually do *better* in time trials while losing most races to B. (How should the example be modified to accomplish this?)

The pairwise best paradox [8]

In the same paper Blythe shows how it can happen that A is favored over both B and C and B is favored over C in two-person races but that in a three-person race C is favored to win. A game involving a single spinner can be constructed as shown in figure 15.5. Reading clockwise from the top, the sectors have probabilities .25, .35, and .40. Each player chooses a track and the spinner is spun. The player with the smallest number wins. If two persons play, the outer track is preferred to either of the others and the middle track is preferred to the inner. If three persons play, the inner track is preferred.

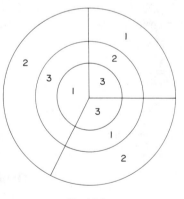

Fig. 15.5

The judge's paradox [37; 17]

In a now famous case in California, a couple was convicted of a crime because testimony from experts showed that the probability was extremely small that any couple would fit all the characteristics of the criminal couple. The accused couple fit the description; therefore, the prosecution argued, they must be the guilty couple. However, this should not be taken as evidence of guilt, for—paradoxically—even though the probability of another couple fitting the description is small, the conditional probability that there is *only one* such couple—given that there is *at least* one such (as eye-witnesses asserted)—need not be small. To compound the paradox, Clark and MacNeil [17] argue that the important computation in establishing guilt beyond reasonable doubt is the conditional probability that the accused are guilty, given both that the accused couple has the characteristics and that the guilty couple has the characteristics.

Conclusion

Statistical procedures generally give correct interpretations of data, and certainly that is what citizens should know and what students should be taught. Large doses of paradox are inappropriate in a class, but a carefully chosen paradox can enliven a class and enlighten students about important statistical issues.

16. Simple Graphical Techniques for Examining Data Geneɪated by Classroom Activities

Carolyn Alexander Maher

CONSIDER a classroom in which the teacher presents a problem ("Which game strategy would produce greater winnings?" "How many rainy days can you expect during your vacation?"), elicits tentative modes of solution from students, encourages simulation activities to confirm or deny these suggestions, and, finally, offers techniques for comparing solutions and suggesting alternatives. Such an atmosphere encourages students to take risks in offering insights, hunches, or beliefs about the resolution of the problem while simultaneously requiring them to evaluate the efficacy of their findings. It permits students to explore alternative solutions without the threat of judgments of right or wrong and lays a foundation for the later, more sophisticated skill of proposing viable hypotheses and testing them.

This article presents several classroom activities that have these learning-liberating characteristics and that introduce students—alone, in pairs, or in groups—to notions of strategies in game theory and Markov chain models. The concepts of random variable and probability distribution and the techniques of graphical summaries and simultaneous comparison should emerge as the problems are resolved.

The four activities described can serve as a paradigm for the invention of suitable classroom activities, a problem faced by every teacher. Selected for their classroom practicability and effectiveness, these activities have been used successfully with above-average junior high school students and average high school and two-year college students.

Class Activity—Measurement

You can try to generate a discussion by posing any or all of the following questions: What are some of the ways in which you (the students) are alike? Different? What are some of the ways in which a few of you are more like each other than others in the group?

A discussion of these questions should lead to the selection of a particular characteristic for study; height is the example presented here. Questions might be raised about the average height of the students, the average height of the girls compared to that of the boys, and so on. The varying guesses should disclose the need for the collection of data and their treatment, presentation, and interpretation.

Table 16.1 gives the data resulting from measuring the height of the nine boys and ten girls, and figure 16.1 presents the data in histogram form.

TABLE 16.1
HEIGHTS (IN CENTIMETERS) OF STUDENTS

| Boys | 157 | 140 | 165 | 150 | 148 | 178 | 132 | 152 | 173 | |
| Girls | 165 | 127 | 135 | 145 | 130 | 135 | 147 | 163 | 150 | 127 |

You might ask your students to compare the heights of boys and girls on the basis of figure 16.1. Other presentations of the same data are offered in the side-by-side histogram in figure 16.2 and the stem-and-leaf plot [136] in figure 16.3. For the stem-and-leaf plot, the numbers within the stem

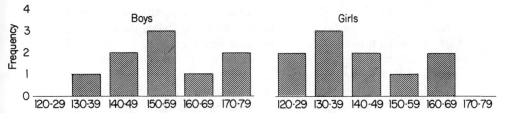

Fig. 16.1. Histograms of the heights (in cm) of boys and girls

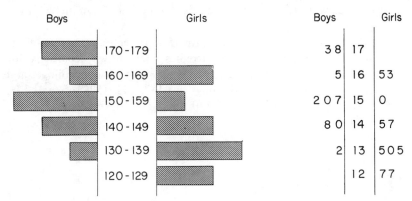

Boys	Girls
3 8	17
5	16	5 3
2 0 7	15	0
8 0	14	5 7
2	13	5 0 5
	12	7 7

<table>
Boys   Girls
170-179
160-169
150-159
140-149
130-139
120-129
</table>

	Boys	Girls
	3 8	17	
	5	16	5 3
	2 0 7	15	0
	8 0	14	5 7
	2	13	5 0 5
		12	7 7

Fig. 16.2. Side-by-side histograms of
heights of boys and girls

Fig. 16.3. Stem-and-leaf plot of
heights of boys and girls

(vertical bar) represent the common left-hand portion of the values re-corded in that line.

You might wish to draw from your students a comparison of the presentations of the same data in figures 16.1, 16.2, and 16.3. Unlike the presentations in figures 16.1 and 16.2, figure 16.3 preserves the individual values while retaining the general shape of a histogram. It thereby presents *all* the data while maintaining the visual impact of figure 16.2.

A discussion of the data might be inaugurated by questions like the following: What is the location of the center of the data for the class? For girls? For boys? Are there unusual (extremely high or extremely low) values? Are the data spread symmetrically about the location of their center?

The answers to these questions involve, along with other observations, a number of calculations—mean, median, and hinges. A box-plot [136] is a graphic presentation of the data, including the results of the calculations and other characterizations, in succinctly synthesized form.

In median-hinge box-plots, the *lower hinge* is defined as the median of the lower half of the data and the *upper hinge* as the median of the upper half. The lower and upper halves of the data are obtained by dividing the ranked data in half at the median value, including the middle value if the number of data values is odd. The *midspread* is the difference between the upper and lower hinges. The *lower extreme* is the smallest and the *upper extreme* the largest of the data values. A long line (more than three times the midspread) between upper hinge and upper extreme (or lower hinge and lower extreme) indicates the presence of an unusually high (or low) value.

Another kind of box-plot is the mean–standard deviation type in which the upper and lower hinges are each one standard deviation from the mean. It, too, provides for graphic comparison of data; however, its structural symmetry excludes its usefulness for the determination of symmetry in the data.

For the data on students' heights, figure 16.4 presents the median-hinge box-plot with medians, hinges, and extremes each drawn separately for boys and girls. The lower extreme, lower hinge, median, upper hinge, and upper extreme for the heights of the boys are 132, 148, 152, 165, and 178, respectively, and for the girls 127, 130, 140, 150, and 165, respectively.

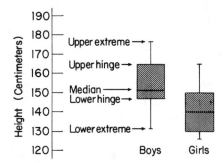

Fig. 16.4. Median-hinge box-plots for boys' and girls' heights

The box-plots show that the median height of the boys is 12 centimeters greater than that of the girls; that the midspread of the girls' heights is 3 centimeters greater than that of the boys (17 centimeters for boys and 20 centimeters for girls); that the middle half of the data on height is symmetric for the girls (the median line is in the middle of the box) and skewed toward larger values for the boys; that the extreme values are neither equidistant from their respective hinges nor from their respective medians for both sexes; and that the shortest boy is slightly taller than the shortest girl, whereas the tallest girl is shorter than one-quarter of the boys.

In short, the most commonly determined characterizations of data are visually present in a single box-plot.

The Random Walk (Single Player) Game

You can introduce this activity by presenting the Random Walk game, which is played on the board illustrated in figure 16.5. Each player has the option of starting from position A or B. Each of eight allowed moves is

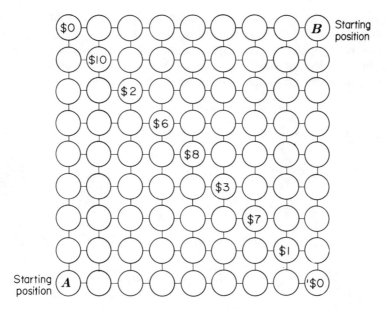

Fig. 16.5. Random Walk game board

determined by the outcome of the roll of a six-sided die. If 1 or 2 appears, the player moves one position horizontally in the direction of the prize diagonal. If 3, 4, 5, or 6 appears, the player moves one position vertically in the direction of the prize diagonal. The last move will always terminate on one of the prize circles on the diagonal.

There is a charge of $6 to play. Four of the nine possible prize winnings equal or exceed the initial charge. Ask your students whether it makes a difference if a player begins at one or the other starting point. Resolving the probable differences of opinion requires a comparison of the outcomes of plays begun from each position—in effect, a simulation exercise. Students might play two games, one from each starting position. The data should be assembled and the strategies compared as in the following simulations from a class of ten students.

Table 16.2 gives the prize winnings for each student for each of the starting positions.

TABLE 16.2
PRIZE WINNINGS (IN DOLLARS) FOR RANDOM WALK GAME

Starting from:	A	10	2	3	6	10	8	6	2	6	0
	B	1	7	3	8	3	8	0	7	3	7

The lower extreme, lower hinge, mean, median, upper hinge, and upper extreme for the data in table 16.2 are 0, 2, 5.3, 6, 8, and 10, respectively, for strategy A and 0, 3, 4.7, 5, 7, and 8 for strategy B.

Figure 16.6 shows a variation of a stem-and-leaf display that employs a six-interval stem: 0–1, 2–3, 4–5, 6–7, 8–9, 10–11, labeled 0, T, F, S, ·, and 1, respectively [136], and the corresponding box-plots. The segments protruding from the sides of the boxes indicate the value of the mean and are placed to show the relative position of the mean and median.

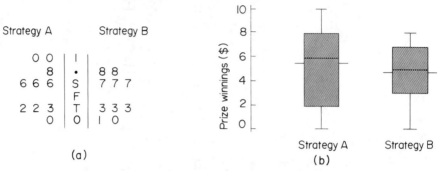

Fig. 16.6. Stem-and-leaf (a) and box-plot with mean (b) displays for prize winnings for strategies A and B

The stem-and-leaf and box-plot displays show that the Random Walk game is not a fair game, since the mean prize winning is less than $6 regardless of whether the player starts at position A or B. Both the mean and median prize winnings are more for strategy A than for strategy B, since a player starting at position A has a greater chance of winning $10.

For more advanced students, the binomial probability distribution can be used to derive a general model for predicting an outcome of the Random Walk game with probabilities assigned to horizontal and vertical moves.

Weather Prediction—a Markov Chain Model

The motivational device for this activity might well be a scenario built around a ten-day vacation. Ask your students to imagine that they've arrived at their hotel, turned on the radio, and heard the meteorologist announce "warm weather with scattered showers throughout the day." They may wonder whether the entire vacation will be spoiled by rain. Perhaps they can find a way of discovering what the number of rainy days during the next nine days might be. Local weather records indicate that a rainy day is followed by another rainy day with probability 2/3 and by a dry day with probability 1/3. A dry day is followed by another dry day with probability 5/6 and by a rainy day with probability 1/6. The arrow diagram in figure 16.7 presents a summary of these conditions.

Assuming that the first day of the ten-day vacation is rainy (R), ask each student to simulate the succeeding nine days employing the transition probabilities given above.

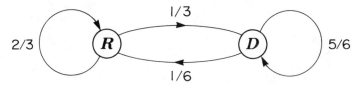

Fig. 16.7. Transition probabilities for Markov chain weather model

The weather of each day can be simulated by tossing a six-sided die nine times. A rainy day succeeds a rainy one if the outcome of the toss is 1, 2, 3, or 4; otherwise (5 or 6), a dry (D) day follows a rainy one. A dry day succeeds a dry one if the outcome of the toss is 1, 2, 3, 4, or 5; otherwise (6), a rainy day follows a dry one.

Figure 16.8 presents a graphical summary of weather data obtained by a class of ten students. The lower extreme, lower hinge, median, mean, upper hinge, and upper extreme are 1, 3, 3.5, 4.4, 7, and 9, respectively.

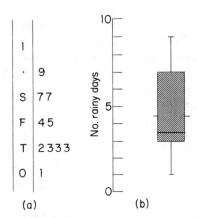

Fig. 16.8. Stem-and-leaf display (a) and median-hinge (with mean) box-plot (b) of number of rainy days during 10-day period

The simulation results and graphical summary indicate that on the average, approximately 4.5 out of 10 days are expected to have rain (actual mean = 4.4) with a 50 percent chance of 3.5 or fewer rainy days (median = 3.5) and a 30 percent chance of 7 or more rainy days (3 observations fall on or above the upper hinge). Thus, more dry days are likely than rainy ones.

Competitive Matrix (Two Player) Game

The matrix game is played by two players, each seeking the maximum possible score. Player A selects one of two rows and Player B one of four columns of a 2 × 4 matrix as shown in figure 16.9. The intersection of a row and a column identifies a score for that play—the score above the diagonal for Player A and below the diagonal for Player B. The winner

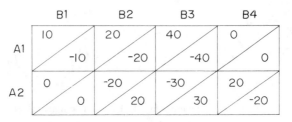

Fig. 16.9. Competitive game score matrix with Player A selections (in rows) and Player B selections (in columns)

receives points at the expense of the loser. Each player's choice is made without the other player's knowledge of it.

You might suggest that students, working in pairs, play several games and see whether they can develop a strategy based on the experience that would result in the highest cumulative scores over many plays of the game. The various student-proposed strategies might then be compared by simulation and display techniques such as those described below.

At first glance, it might appear that Player A should always select row A1, for it suggests greater winnings. This underestimates Player B, who would soon recognize that the appropriate response is B4, except that these result in neither player's winning. Player A must therefore seek to outmaneuver Player B by varying his or her own choices.

One such strategy for Player A might be to select row A1 with probability 2/3 and A2 with probability 1/3 (this strategy maximizes Player A's minimum expected gain). A strategy for Player B might be to select B3 with probability 2/3 and B4 with probability 1/3.

To implement this strategy A1 might use three cards, with A1 on two and A2 on one. The cards should be thoroughly shuffled and one selected at random to identify the choice for each play.

The objective is to compare the strategies using the total cumulative score obtained by each student after, say, ten plays of the game. Figure 16.10 gives the stem-and-leaf and box-plot (with mean) displays for such a classroom simulation activity. Notice that the stem-and-leaf display contains stem intervals of 10 up to 100 and intervals of 100 thereafter, with all scores in multiples of 10. The lower extreme, lower hinge, mean, median, upper hinge, and upper extreme are 10, 50, 131, 150, 170, and 270 for strategy 1 and 40, 50, 68, 70, 80, and 90 for strategy 2.

Both displays suggest that from the point of view of Player B, strategy 2 is preferable to strategy 1. It results in fewer points for Player A.

Following the simulations and graphical comparisons of some student-suggested Player A and Player B strategies, a further classroom activity might be to compare two or more of the suggested Player B strategies against the Player A strategy described in the examples above. This activity should suggest to students that the Player A strategy, that is, selecting

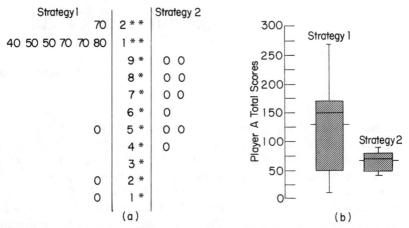

Fig. 16.10. Stem-and-leaf (a) and box-plot with mean (b) displays for Player A total scores for strategies 1 and 2

A1 with probability 2/3 and A2 with probability 1/3, is a reasonably good strategy for Player A.

For more advanced students, you may wish to discuss such concepts as expected value and the relationship of matrix game theory to linear programming. In fact, these techniques could be used to show that the strategy for Player A described above employed against the optimum strategy for Player B will maximize A's average score [120].

Conclusion

The four activities described here demonstrate that focusing on the structure and content of material is compatible with an organization of activities designed to excite the curiosity and enthusiasm of the learner. Ranging from the relatively simple graphical data summaries, as in the height example, to those that may suggest hypotheses for confirmation or denial by data-analysis techniques, as in the matrix game, the approach is adaptable to differing backgrounds and abilities. One might even extend the procedures to include techniques based on the construction of a confidence interval for each median [79].

It should be noted that sample sizes used in these presentations are small for illustrative purposes, but it is suggested that samples in a classroom setting be at least twenty-five to augment confidence in the comparisons.

The techniques presented here are those employed by practicing researchers. Although exploratory in nature, they are intended as a first step to reveal clues and suggest hypotheses that in turn can be confirmed or denied on the basis of appropriate data-analysis techniques.

The particular activities offered here are intended merely to serve as examples. No doubt you and your students will suggest numerous others.

17. Random Digits and the Programmable Calculator

Lennart Råde

Calculators and computers have become important in the study of mathematics, both for developing content areas and for their applications. For example, computers are used with great success in number theory. Also, the recent computer proof of the four-color theorem shows that the computer is important for pure mathematical research. This is reason enough to introduce the programming of calculators and computers in the teaching of mathematics at the school level. But an even more important reason is that the use of calculators and computers can stimulate the teaching of mathematics by offering the possibility for a concrete and challenging approach to the subject.

Probability theory is one area of mathematics whose teaching may be significantly influenced by the use of programmable machines. Programmable calculators are especially useful here, since they provide students with an easily available capacity for numerical calculations and for simulations of random phenomena.

Simulation has always been important in the application and teaching of probability theory. For example, the study of relative frequencies in connection with the tossing of coins and dice has long been a standard technique for acquainting students with the practical background of the concepts of probability. However, one difficulty has been that such simulations take a long time to perform and thus become tedious for the students. It is not exciting to toss a coin 1000 times!

Now, however, with the programmable calculator, we have an efficient and motivating tool that can be used to simulate random experiments. Thus simulations can become an essential part of the teaching of probability and statistics.

It is true that programmable calculators are slow compared to computers, but from a didactic point of view this can be an advantage. Pauses in a simulation program make it possible for a student to follow what happens in a random experiment. This makes it possible to develop in a student's mind a concrete understanding of random variation, of the law of large numbers, of what it means for an experiment to have no expectation, and so on. Simulations also give concrete background for theoretical investigations in probability and statistics and motivate the student to engage in such investigations.

Following are some concrete examples of how the programmable calculator can be used in the teaching of probability and statistics at the school level. The examples are based on experiences from an experimental course in probability and statistics for the Swedish Secondary Schools [99] and from experimental coursework in the Elements of Mathematics Series of the Comprehensive School Mathematics Program [101].

Generation of Random Digits

Here is a very simple method for generating a sequence of random digits, simulating a sequence of outcomes obtained by spinning a fair decimal spinner (figure 17.1).

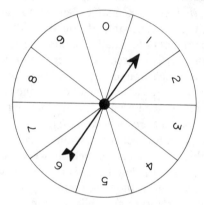

Fig. 17.1

Start with a decimal number x_0 between 0 and 1 that has at least five decimal places (use as many places as the calculator allows); for example,

$$x_0 = 0.379645937.$$

Then generate a new decimal number between 0 and 1 by multiplying x_0 by 147 and taking the fractional part of the product. If, in general, the fractional part of x_0 is denoted by FRAC(x), the new decimal number x_1 is given by the formula

$$x_1 = \text{FRAC}(147x_0).$$

Thus from our example we obtain

$$x_1 = 0.807952740,$$

since $147x_0 = 55.80795274$. The first decimal place of x_1 is the first random digit. Here it is 8.

Next, apply the same procedure to x_1; that is, calculate

$$x_2 = \text{FRAC}(147x_1),$$

which in this case gives $x_2 = 0.769052800$. Thus, the next random digit is 7. Continue according to the formula

$$x_{n+1} = \text{FRAC}(147x_n),$$

and each time record the digit in the first decimal place. This method of generating random digits is called the *147-generator*.

In this example we obtain, at the start, the following random digits:

87043	41804	05311	40273	27915
53142	65953	55217	63283	57194
39709	83188	99544	65505	07599
60997	35524	73894	40729	00495
28036	55017	12654	26233	08594

Random digits are extremely important and useful in applied probability. A method of generating random digits on a calculator or computer is called a *random digit generator*. The 147-generator is just one such example. Random digit generators are, of course, really not random in the same way as the experiments of actually spinning a spinner or tossing a die. When the first number x_0 is chosen, all other numbers are decided on as well. However, it has been found that the 147-generator gives results similar to those obtained with a spinner like the one shown in figure 17.1. Sometimes random digits generated by a random digit generator such as the 147-generator are called *pseudorandom digits*.

The students will probably be curious about the multiplier 147. They may want to try other multipliers as well; however, they should wait to make such investigations until they have learned what it means to check a random generator (discussed in the next section). Experience indicates that 147 is a suitable multiplier. Other multipliers recommended in the literature are 83, 117, 123, 133, 163, 173, 187, and 197.

Observe that the sequence $x_0, x_1, x_2, \ldots$ is periodic. When a number x_n is repeated, the next numbers will be repeated as well. Because a finite number of possibilities for x_n exist, sooner or later a number already obtained will reappear. By using the multipliers above, periodic sequences with periods of maximum length will be obtained. (I have found that 137 also gives satisfactory results, even if it does not give maximum periods.) It is, of course, desirable to have a random digit generator with a long period. In particular, the period should be longer than the number of random digits required in the simulation.

The numbers $x_0, x_1, x_2, \ldots$ used in the random digit generators described above also have a random character. Let us call them "random numbers between 0 and 1" and describe them as results of successive random choices of points between 0 and 1 on the number line. A detailed discussion of random digit generators is found in [100].

Several activities with random digits are suitable for students to undertake at this introductory stage. They can, for example, write a program that will generate random digits as above, limiting the display to five digits each time. Or they can write a program that will generate 1000 random digits and find the number of odd digits generated.

Testing a Random Digit Generator

It is important to give students confidence in the random digit generator they are using, whether this is the 147-generator or some other. One simple method is to use the frequency test, that is, to find the frequencies of the ten different digits in a generated sequence of digits. For a good random digit generator, these frequencies should be about the same. As an alternative they can calculate the relative frequencies, all of which should be about 0.1. Writing a program for a frequency test will also give the students experience in devising sorting techniques. However, the following discussion shows that a frequency test is not enough.

Suppose "random digits" are generated as follows: first, 100 0s are generated, then 100 1s, then 100 2s, and so on. Obviously this is not a good random digit generator, but it will pass the frequency test. A more sophisticated test is the poker test, which is made as follows. Generate 400 random digits, for example. Group them in order in 100 groups of four digits in each group. These four-digit groups can be classified according to figure 17.2, which also gives probabilities for different outcomes with the assumption that the random digit generator is a perfect one. A good random digit generator should give these outcomes with relative frequencies close to the probabilities in the figure.

Outcome	Probability
All different, e.g., 7093	0.504
One pair, e.g., 1731	0.432
Two pairs, e.g., 1771	0.027
Three of a kind, e.g., 8388	0.036
Four of a kind, e.g., 5555	0.001

Fig. 17.2

Students can perform a poker test on their calculators in different ways. They can generate four digits at a time and classify the outcomes manually. Or, they can write a program that will have the calculator make a complete poker test.

Observe that calculating the probabilities that are used in the poker test is a good exercise in computing probabilities. But more important for the students might be that they meet an application where the calculation of probabilities is needed. The actual calculation of the probabilities can be done in different ways depending on what approach to probabilities is used. With a traditional combinatorial approach, the probabilities can be found as indicated in figure 17.3. The easiest way to find the probability for two pairs is to subtract the sum of the probabilities from 1.

Outcome	Probability
All different	$\dfrac{10 \cdot 9 \cdot 8 \cdot 7}{10^4}$
One pair	$\dfrac{\binom{4}{2} \cdot 10 \cdot 9 \cdot 8}{10^4}$
Three of a kind	$\dfrac{4 \cdot 10 \cdot 9}{10^4}$
Four of a kind	$\dfrac{10}{10^4}$

Fig. 17.3

If a tree approach is used, these probabilities are easily found from appropriate tree diagrams. A tree approach to probability is used in [33] and [34].

Introduction of Expectation

An outline will follow showing how the concept of the *expectation* of a random experiment may be introduced when the students are working with programmable calculators. (*Note:* With x between 0 and 1, $6x$ is between 0 and 6 so that $\text{INT}(6x) + 1$ is a whole number from 1 to 6—that is, the result of rolling a die.)

First a random experiment is simulated according to the flowchart in figure 17.4. Successive tosses of a symmetric die are simulated, and after each toss the average of all the outcomes so far is calculated. For this situation it is suitable to have the calculator round off the averages to the nearest tenth. As the program runs, it turns out that the average stabilizes to the number 3.5. How can this be explained?

Suppose that the frequencies of the outcomes 1, 2, 3, 4, 5, and 6 in n trials are n_1, n_2, n_3, n_4, n_5, and n_6. Then the average $\overline{x}$ in these trials is given by

$$\overline{x} = \frac{n_1 \cdot 1 + n_2 \cdot 2 + n_3 \cdot 3 + n_4 \cdot 4 + n_5 \cdot 5 + n_6 \cdot 6}{n}.$$

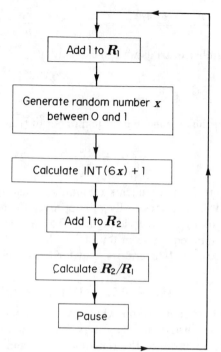

Fig. 17.4

This can be expressed as

$$\bar{x} = \frac{n_1}{n} \cdot 1 + \frac{n_2}{n} \cdot 2 + \frac{n_3}{n} \cdot 3 + \frac{n_4}{n} \cdot 4 + \frac{n_5}{n} \cdot 5 + \frac{n_6}{n} \cdot 6.$$

But in a large number of trials the relative frequencies n_i/n can be expected to be close to 1/6. That is, the average will be close to

$$\frac{1}{6} \cdot 1 + \frac{1}{6} \cdot 2 + \frac{1}{6} \cdot 3 + \frac{1}{6} \cdot 4 + \frac{1}{6} \cdot 5 + \frac{1}{6} \cdot 6 = 3.5.$$

The number 3.5 is called the *expectation* when a symmetric die is tossed.

Let us consider another random experiment, namely, to spin the spinner in figure 17.5 and to calculate the average after each spin as in the previous example. But here we shall make a mathematical analysis before we perform the experiment. Let n_0, n_1, and n_2 be the frequencies of the outcome 0, 1, and 2, respectively, in n trials. Then the average $\bar{x}$ will be

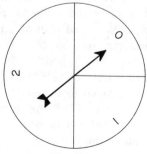

Fig. 17.5

$$\bar{x} = \frac{n_0 \cdot 0 + n_1 \cdot 1 + n_2 \cdot 2}{n} = \frac{n_0}{n} \cdot 0 + \frac{n_1}{n} \cdot 1 + \frac{n_2}{n} \cdot 2.$$

But in a large number of trials we can expect that

$$\frac{n_0}{n} \approx \frac{1}{4}, \quad \frac{n_1}{n} \approx \frac{1}{4}, \quad \text{and} \quad \frac{n_2}{n} \approx \frac{1}{2}.$$

Thus we can expect the average in a large number of trials to be close to

$$\frac{1}{4} \cdot 0 + \frac{1}{4} \cdot 1 + \frac{1}{2} \cdot 2 = 1.25.$$

It is now challenging for the students to simulate this experiment on their calculators to see if the averages really will approach the number 1.25.

The simulation of the spinner in figure 17.5 can be made as follows. Generate random numbers between 0 and 1 with a random number generator—for example, the 147-generator—and for each such random number, x_n, calculate

$$Y_n = \text{INT}(x_n + 0.5) + \text{INT}(x_n + 0.75).$$

It is not difficult to understand why this will simulate the spinner in figure 17.5. Consider where in the interval from 0 to 1 a number x_n should be in order to produce the outcomes 0, 1, and 2, respectively. If $0 \leq x_n < 0.25$, $y_n = 0$; if $0.25 \leq x_n < 0.5$, $y_n = 1$; if $0.5 \leq x_n \leq 1$, $y_n = 2$.

These experiments prepare the students for the introduction of the expectation as a weighted average of the outcomes with the corresponding probabilities as weights. They will also, from the beginning, give students an understanding of the expectation as a prediction of the average in a large number of trials.

Now several random experiments can be simulated and the average obtained in a large number of trials. This is used to estimate the expectation when it is difficult to calculate or to study the random variation of the averages around the expectation.

Figure 17.6 illustrates an interesting random experiment. Two towers are built of m and n blocks, respectively. A tower is chosen at random and a block is moved from this tower and put on top of the other tower. The procedure is continued until all the blocks have been moved from one tower to the other. Find the expected number of moves necessary to produce this result.

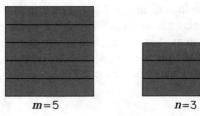

$m=5$ $n=3$

Fig. 17.6

The flowchart in figure 17.7 describes how this experiment can be simulated on a programmable calculator. It is easy to extend the program so that a large number of trials (100 or 1000, for example) are performed and the average number of moves is calculated.

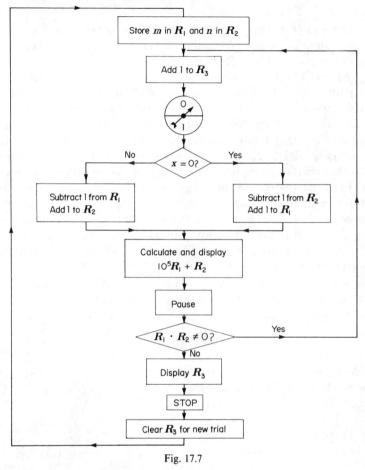

Fig. 17.7

It is not easy at the school level to derive a simple expression for the expectation in this case. But this expression may be discovered by the students from the results of their simulations. (The expectation is $m \cdot n$. The random experiment is equivalent to a classical probability problem dealing with the duration of a fair game between two players with capitals m and n, respectively.)

Many other challenging situations can be studied in a similar way—for example, various random experiments dealing with the drawing of elements at random and without replacement from a population of elements of two kinds. Several such simulations are discussed in [100].

18. Correlation, Junior Varsity Style

Annette N. Matsumoto

STATISTICS and probability are often neglected topics in seventh- and eighth-grade mathematics courses. When they are covered, popular topics include graphs, averages, and simple probability. Yet other aspects of statistics and probability affect us all in our daily lives—among them correlation.

The junior high school level offers an excellent opportunity to introduce intuitive notions without the detailed calculations often associated with statistics and probability. What follows is a section of a unit on statistics with probability that I have written and used in my eighth-grade class at the University of Hawaii Laboratory School. This section on correlation deals with linear relationships, with many of the examples being clearly linear. Exposing students to the notion of correlation is the primary objective. Curvilinear relationships can be discussed informally after completion of the section, just to make students aware of other kinds of relationships.

I have found that the subsection dealing with the "cause-effect" fallacy leads to interesting discussions and the sharing of television commercials that typify this fallacy.

It is hoped that the following lesson will give junior high school mathematics teachers a hint of how high-powered statistical topics can be handled at an intuitive level!

Correlation

A detective who saw a large footprint at the scene of a crime would probably look for a very tall suspect. The detective would be using the relationship that people with large feet are usually very tall.

Relationships like this are very useful because changes in the value of one condition (for example, height) are associated with changes in the value of another (foot size). Thus it is possible to *predict* the value of one condition if the value of the other is known.

The content of this lesson is now a part of *Statistics with Probability*, a unit of the Intermediate School Mathematics Program of the University of Hawaii, College of Education.

126

Once a relationship is suspected, statisticians look into the strength and the nature of the relationship. We say that they are looking for the *correlation* between two conditions, or variables.

One kind of relationship is that in which both variables increase or both decrease. The two variables are then said to have *positive* correlation. The example given above shows two variables that are correlated positively. Both variables, height and foot size, either increase or decrease. Thus, a man with size 13 shoes probably would not be under six feet tall! Of course, there may be exceptions.

Another type of relationship is one in which one variable increases as the other decreases. The variables are then said to have *negative* correlation. As an example, let us look at the relationship between the selling price and the amount of goods people will buy. The higher the price, the less people tend to buy. One year, because of crop failure,

COFFEE
$8 LB.

WHAT? GUESS I'LL BE DRINKING HOT WATER!

SALE!
SOFTEST BRAND TISSUES
3 boxes for $1
reg. price 79¢ ea.

BOY, THAT'S SO CHEAP, I MAY COME BACK TOMORROW!

the price of coffee went up to about $8 a can. People began buying tea instead of coffee. By the same token, when prices are low—when there is a sale, for instance—people will buy more than they usually would.

The third "relationship" is that no relationship exists. An example is the relationship between a person's height and month of birth.

One can make rather detailed calculations in statistics to determine whether two variables are positively or negatively correlated or not correlated at all, as well as how strong the correlation is. However, we shall be using a simpler method, which gives a rough idea. This method is the scatter diagram.

The Scatter Diagram

A scatter diagram is a graph of two variables. For each pair of variables, state whether you would expect a positive correlation, a negative correlation, or no correlation.

1. Ages of husbands and their wives
2. Level of education and income
3. Belt size and sense of humor

4. Outside temperature and gallons of water used

5. Speed of travel and travel time

6. Bowling average and spelling test score

7. Height and weight

8. Amount of homework done and score on test

Positive correlation

Let us look at an example of two variables with positive correlation.

Example: Temperature and rate of cricket chirps

Table 18.1 shows different temperatures and the corresponding number of times a cricket chirps a minute. Figure 18.1 shows a graph of these values—a scatter diagram. Notice that the temperatures are on the *horizontal* or *x-axis* and the rate of chirps on the *vertical* or *y-axis*. Thus, temperature is the *x-variable* and the rate of chirps is the *y-variable*. The points are plotted as follows: each pair of values is thought of as an ordered pair. For example, (18°C, 110) represents the temperature at 18°C and its corresponding rate of 110 chirps a minute. You will notice that the graph shows

TABLE 18.1

TEMPERATURE AND RATE OF
CRICKET CHIRPS

Temperature (°C)	Rate of Chirps (per minute)
18	110
19	110
20	130
21	130
21	135
23	154
24	158
24	157
24	169
26	176
26	180
26	179
29	201
29	209
31	210
32	230

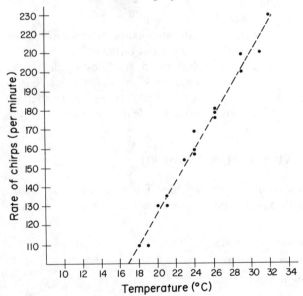

Fig. 18.1. Temperature and cricket chirps

the trend of the points rising as we look from left to right. In fact, these points appear to be in almost a straight line!

Thus, variables that are positively correlated show a trend rising from left to right. The closer the points get to a straight line, the stronger the relationship is. All the scatter diagrams in figure 18.2 indicate a positive correlation between the *x*- and *y*-variables.

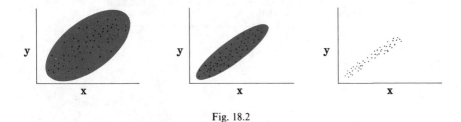

Fig. 18.2

Negative correlation

Figure 18.3 is a scatter diagram showing negative correlation. It shows the latitude and average yearly temperature of twenty cities in the United States. Notice that the graph shows the trend of points falling as we look from left to right. That is, the higher the latitude, the lower the average temperature.

Can you predict what a scatter diagram would look like when there is no correlation? (Think before you answer!)

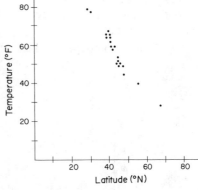

Fig. 18.3. Latitude versus average yearly temperature of twenty U.S. cities

Activity: Making Scatter Diagrams

In this activity, you will collect data on three variables: *(a)* month of birth, *(b)* length of foot, and *(c)* length of forearm. Then you will combine the class data to make two scatter diagrams.

Part 1: Collecting the data

a) *What is your month of birth?*

If you were born in January, record "1" in your notebook; if you were born in February, record "2," and so on.

b) *What is the length of your foot?*

Measure the length of your foot to the nearest centimeter. Record the measurement in your notebook.

c) *What is the length of your forearm?*

Measure from the inside bend of your elbow to your wrist (to the nearest centimeter). Record the measurement in your notebook.

Part 2. Organizing the data

Record class results on a table like the one in table 18.2.

TABLE 18.2

Student	Month of birth (a)	Length of foot (b)	Length of forearm (c)
1			
2			
3			
4			
5			
6			
.			
.			
.			

Part 3. Scatter diagram #1

Is there any correlation between a person's month of birth and the length of his or her foot?

a) Using graph paper, make a grid like the one shown in figure 18.4.

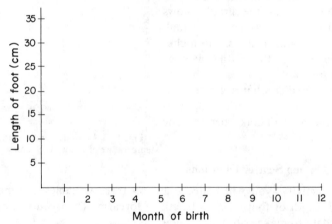

Fig. 18.4

b) Using data from columns *a* and *b* of table 18.2, plot each ordered pair as you do rectangular coordinates. For example, if student 1 was born in February (2) and has a foot length of 24 cm, locate the point (2, 24 cm) on your graph and put a dot there, as in figure 18.5. Continue doing this until you have a point on your scatter diagram for every student in your class. (If two or more ordered pairs are identical, put an *x* at that point.)

c) Does there appear to be any correlation between the two variables (month of birth and foot length)?

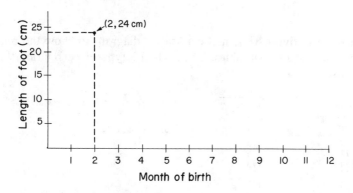

Fig. 18.5

Part 4. Scatter diagram #2

Is there any correlation between the length of a person's foot and fore-arm?

a) Make a grid.

b) Using columns *b* and *c* of your table, plot each ordered pair. For example, if student 1 has a foot length of 24 cm and a forearm length of 23 cm, locate the point (24 cm, 23 cm) on your graph.

c) Does there appear to be any correlation?

Exercise 1

For each of the diagrams in figure 18.6, tell the type of correlation that exists between the two variables *(positive, negative,* or *none),* and describe the strength of the correlation *(very strong, strong, weak).*

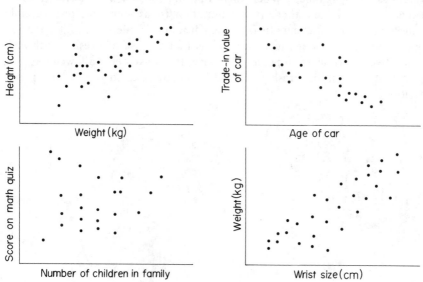

Fig. 18.6

Exercise 2

For each table in figure 18.7, make a scatter diagram to show the relationship between the two variables. Then tell the type of correlation that appears to exist.

Male Student's Height (cm)	Father's Height (cm)
178	175
177	175
169	173
176	170
163	163
180	179
173	163
173	169
188	193
170	188
175	170
168	165
201	193
165	163
173	180
173	173
160	152

Average Monthly Temperature	
Washington, D.C. (°F)	Melbourne, Australia (°F)
44	78
45	78
55	75
65	68
75	62
83	57
86	56
84	59
78	63
67	67
56	71
45	75

Fig. 18.7

A Caution about Correlation

As consumers, we are faced with correlations in advertising and in public relations "propaganda." When there appears to be a high correlation between two variables, there is a temptation to assume that one variable causes the other to occur. In fact, the consumer is allowed to interpret exactly that! The message in figure 18.8 is that the use of Fresh mouthwash is positively correlated to popularity. Also, the consumer is allowed to interpret that the use of Fresh mouthwash actually caused the sudden popularity!

Fig. 18.8

High correlation alone does not necessarily mean that one variable causes the other—it just indicates a possibility. Sometimes the two variables are influenced by a hidden third variable. For example, it was observed that in certain communities there was a high positive correlation between ministers' salaries and liquor consumption. As the ministers' salaries rose, so did the amount of liquor consumed in the community. Could it be that as their salaries rose, the ministers were encouraging liquor consumption? Or could the income from liquor sales be going to the ministers' salaries?

Actually, the cause of both could be explained by the upward economic trend of the entire community. Under prosperous conditions, everyone's salary goes up and people buy more than they would under "tight money" situations.

It is unlikely that anyone would believe that ministers' salaries and liquor consumption are causally related. However, high correlations between other variables are often assumed to be causally related. For example, many people believe that a lack of education causes involvement in crime.

Thus, high correlation alone does not mean that one variable causes the other. To find out, experiments or studies must be carefully conducted.

Project

Make a list of television commercials that can be interpreted to say that using the product causes something "good" to occur. Discuss what other factors might cause that "something good" to occur.

Predicting from graphs

Let us look again at the scatter diagram in figure 18.1 comparing cricket chirps. As we have seen, the points of the diagram appear to be in almost a straight line. Notice that when we draw in the line, as shown in the figure, we can predict the rate a cricket will chirp when the temperature is, say, 34°C.

Graphs that seem to show a distinct trend are often used to make predictions. Although such predictions may later prove to be incorrect, they are at least reasonable estimates, not wild guesses!

Activities

1. Figure 18.9 is a line graph showing Hawaii's de facto population (actual population, excluding residents temporarily absent and including visitors present) from 1965 to 1978. Assuming the growth rate remains the same, predict what year the population will reach 1 million.

2. Figure 18.10 is a graph showing the record-setting times for the mile run from 1884 to 1979. Notice that the points lie almost in a straight line. The dashed line is an *estimate* of that straight line. It can be used to predict what might occur in the future.

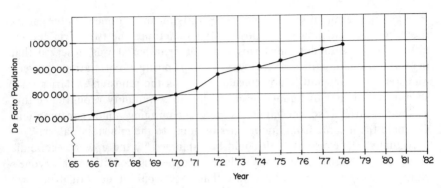

Fig. 18.9

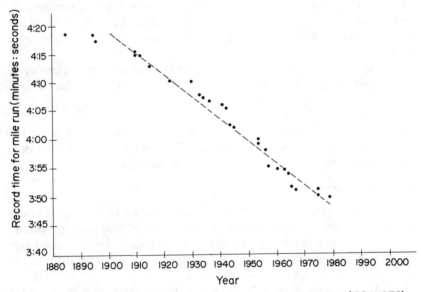

Evolution of the Record-setting Times for the Mile Run (1884-1979)

Fig. 18.10

a) Using the dashed line, predict when the mile will be run under 3 minutes, 45 seconds.

b) Using the dashed line, predict the record time for the mile in 1984.

c) When do you predict the mile will be run in 3 minutes, 40 seconds?

d) Do you think it is reasonable to predict when the mile will be run in 2 minutes? Why, or why not?

19. An Area Model for Solving Probability Problems

Richard D. Armstrong

Most American children have an intuitive concept of randomness, partially due to playing games involving dice, spinners, and cards. Since probability provides a rich source of problem-solving experiences, we decided to extend our students' informal experiences and include probability as an integral part of our elementary mathematics curriculum. We developed stories and games for the second and third grades that introduce such concepts as expected frequency, equally likely events, and prediction. The students' reactions to these activities indicated to us their capability of progressing to the analysis of one-stage probability experiments through combinatorial methods. In one third-grade class the students considered the thirty-six equally likely outcomes when two dice are thrown and determined that there are six ways for a sum of seven to occur. Thereby they calculated that the probability of rolling a sum of seven is 6/36, or 1/6.

Their success and enjoyment in analyzing several one-stage probability situations demonstrated that these students were capable of considering more complex, multistage experiments in the intermediate grades. However, traditional arithmetic solution techniques of such problems tend to require either unwieldy combinatorial analysis or a well-developed understanding of the addition and multiplication of fractions. Of course, the consideration of these problems could be postponed to later grades, but even for more mature students the computational aspects of arithmetic solutions often tend to obscure rather than illuminate the underlying probabilistic concepts.

The need for an alternative model for solving probability problems became apparent. To be appropriate for students in the intermediate grades, we thought the model should—

- be sufficiently powerful to handle fairly sophisticated probability problems;
- rely primarily on mathematical skills that the students have already acquired;
- be consistent with the students' current understanding of probabilistic concepts;

The activities described in this essay were taught to students in the St. Louis area as part of the development of the Comprehensive School Mathematics Program at CEMREL, Inc., St. Louis, Missouri, under a contract from the National Institute of Education, Department of Health, Education and Welfare. The opinions expressed do not necessarily reflect the position of the National Institute of Education, and no official endorsement should be inferred.

135

• support the eventual development of more advanced solution techniques.

Considering that most probability situations intrinsically involve fractions and that a common model for fractions involves the partitioning of circular or rectangular regions ("pies" or "cakes"), perhaps it is natural that we developed a geometric model to satisfy the criteria above. In this model, a unit square is divided into regions so that the areas of the regions are proportional to the probabilities involved in the situation. The following three activities indicate the use and development of this model and moreover illustrate its pedagogical and mathematical characteristics. (The first two activities were presented to two intermediate school classes, grades 5–6, in the St. Louis area. The third activity is proposed for junior high school students.)

Marriage by Chance

Mr. Simons, a fifth-grade teacher, tapes a poster (fig. 19.1) on the board and with appropriate embellishment tells the following story, occasionally allowing the students to react and comment. (This story is inspired by a popular short story, "The Lady or the Tiger?" by Frank Stockton [129].)

"The king and queen of a medieval kingdom arranged a marriage for their daughter to Prince Cuthbert from a neighboring kingdom. The princess accepted this plan without enthusiasm. A short time before the proposed wedding day she met Reynaldo—handsome, clever, romantic, but only a peasant. Their love developed quickly and secretly, but inevitably the king learned of their relationship. Irate, he ordered that Reynaldo be thrown into a room full of tigers. But in response to his daughter's pleas, he offered a compromise: Reynaldo would walk through a maze, each path leading to one of two rooms. While the hungry tigers waited in one room, the hopeful princess would wait in the other room. If Reynaldo entered the latter room, he and the princess could marry."

Pointing to the poster, Mr. Simons continues, "The king showed the princess a map like this one of the maze and let her decide in which room to wait. Remember that Reynaldo does not have a copy of the map and can only guess which paths to follow. Which room is he more likely to enter, A or B?"

Some students suggest that Reynaldo's probability for entering each room is 3/6, or

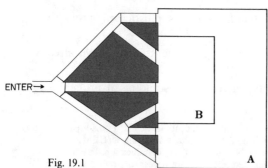

Fig. 19.1

1/2, because there are three doors into each room. However, other students realize that the answer is not so obvious, since Reynaldo is more likely to arrive at the third door from the top than at other doors because there is a path that leads directly from the entrance to the third door. After more discussion, the majority of the class votes that they think the princess should wait in Room B.

Mr. Simons draws a large square on the board and suggests, "Let's use this square to determine the probability that Reynaldo will enter Room B. When he enters the maze, what is the first choice Reynaldo must make?" When a student responds that Reynaldo must choose to take the upper path, the middle path, or the lower path, Mr. Simons adds some information to the square. See figure 19.2(a).

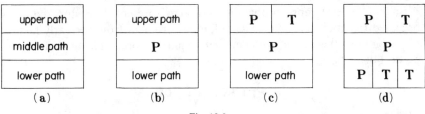

Fig. 19.2

"I've divided the square into three equal parts, since each of the three paths is equally likely to be chosen by Reynaldo," explains Mr. Simons. "What happens if Reynaldo chooses the middle path?"

"He's lucky and walks straight to the room where the princess is waiting," observes a student. Mr. Simons shows this by marking "P" in the center section of the square, as in figure 19.2(b). He continues, "What happens if Reynaldo chooses the upper path?" Observing that the upper path splits into two paths, the students state that Reynaldo's chances of reaching each room would then be the same. They agree to indicate this by dividing in half the part of the square labeled "upper path," as in figure 19.2(c). Then the class correctly divides and labels the region for the lower path; see figure 19.2(d).

For contrast, Mr. Simons colors gray the regions marked "P" and red the regions marked "T," as in figure 19.3(a).

Looking at the square convinces the class that

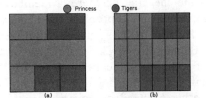

Fig. 19.3

they have placed the princess in the correct room, since more than half the square is colored gray. Mr. Simons agrees and inquires how they could calculate exactly Reynaldo's probability of finding the princess. With hints and encouragement, the class decides to divide the square into small pieces all the same size and to count the number of gray pieces and red pieces. See figure 19.3(b). Out of eighteen pieces of the same size, eleven are gray and seven are red. Therefore, Reynaldo has eleven out of eighteen chances to find the princess. His probability of success is 11/18, or almost 2/3. His success is not guaranteed, of course, but the class did place the princess in the better room.

Some students at first were intent on finding clever ways for Reynaldo to detect and avoid the tigers. Rather than being out of place, this humorous diversion emphasized the need to accept certain restrictions when a situation is being modeled. As in real-life applications, the situation had to be idealized. An advantage of embedding a problem in a story instead of using a real example is that the necessary restrictions can be minimized and well controlled.

Solving several more probability problems presented in story contexts prepared the students to consider a famous problem from the early history of probability theory—a problem that requires more sophisticated mathematical insights.

A Problem of Points

In the history of mathematics, the first probability questions arose from games of chance. One particularly intriguing problem, now called the "problem of points," appeared as early as the fourteenth century. The following is an example of the problem. Two gamblers play a game for a stake that goes to the first player to gain ten points. If the game is stopped when the score is 9 to 8, how should the stake be divided between the two players? It is assumed that the players have equal chances of winning each point.

This problem was popular and controversial in Europe in the sixteenth and early seventeenth centuries. In 1556, Tartaglia claimed to have the solution but simultaneously declared that any solution is "judicial rather than mathematical," that is, it must be agreed on by the two players (an astute commentary on applied mathematics!).

In 1654 the chevalier de Méré, a member of King Louis XIV's court in France, encountered the problem through his interest in mathematics and gambling. He proposed the unsolved problem to a young French mathematician, Blaise Pascal. The ensuing correspondence between Pascal and his older friend, Pierre de Fermat, reveals that they developed three distinct techniques for solving this problem. The application of these techniques to other probabilistic questions provided an impetus to mathematicians and eventually led to the development of modern probability theory.

Embedding the problem of points into a children's game and using the area technique allows students in the intermediate grades to solve this historically significant problem. (The Belgian mathematics educators

Frédérique and Georges Papy discovered this solution technique for the "problem of points." Their solution revealed to our staff the potential of the method in many other situations.)

Let's listen to Ms. Kell as she describes a game to her class. "Rita and Bruce play a game. Rita has one red marble and one blue marble. With her hands behind her back, she mixes them and then puts one marble in each hand. Bruce chooses a hand. If he selects the hand with the blue marble, he scores one point. Otherwise Rita scores one point. The procedure is repeated, and the winner is the first player to score ten points."

After playing the game a few times in class, Ms. Kell suggests the following situation. "One day, Rita and Bruce must stop a game when the score is Rita 9 and Bruce 8. (See fig. 19.4.) If they continue the game the next day, what is the probability that Rita will win?"

After discussing the game and making some estimates, the students use a square to analyze the situation. If the score is 9-8, the next score will be either 10-8 or 9-9, with equal likelihood. Divide the square into halves, as in figure 19.5(a). Rita wins if the score is 10-8. Color the appropriate region red for Rita, as in figure 19.5(b). If the score reaches 9-9, the game is fair. Color half the appropriate region red for Rita and half gray for Bruce. See figure 19.5(c).

Three-fourths of the square is colored red and one-fourth gray. Therefore, when Rita is leading 9 to 8, the probability of her winning is 3/4 and the probability of Bruce winning is 1/4. Because of the symmetry induced by using one red marble and one blue marble, we can immediately conclude that if Bruce were leading 9 to 8, his probability of winning would be 3/4 and Rita's probability of winning would be 1/4.

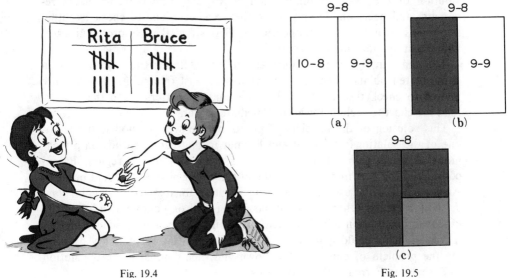

Fig. 19.4   Fig. 19.5

The solution for the problem when Rita is leading 9 to 7 is similar and reveals a useful shortcut. If the score is 9-7, the next score will be either 10-7 or 9-8, with equal likelihood. See figure 19.6(a). Rita wins if the score is 10-7; see figure 19.6(b). For the intermediate score 9-8, we could consider the scores 10-8 and 9-9. But the previous argument shows that if Rita leads 9-8, her probability of winning is 3/4. Therefore the region for "9-8" can immediately be colored three-fourths red and one-fourth gray. Once the square is divided into regions of the same size, there are seven red pieces and one gray piece. Therefore, if Rita is leading 9 to 7, her probability of winning the game is 7/8. See figure 19.6(c).

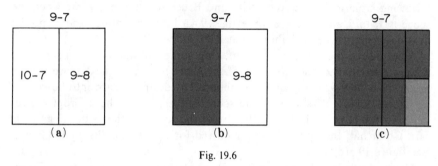

Fig. 19.6

By applying the area technique, students can now determine Rita's and Bruce's winning probabilities for any intermediate score. Such a task appears uninteresting and tedious. However, the use of the area technique has suggested a very natural application of the concept of recursion—for example, using the computed result for the "9-8" problem to shorten the solution of the "9-7" problem. In fact, by detecting patterns, using recursion, and occasionally employing the area technique to check hypotheses, a class is able to determine quickly Rita's and Bruce's probabilities of winning at any intermediate score.

The chart in figure 19.7 indicates the odds (Bruce:Rita) for winning a game to ten points when each player has at least five points. Readers are invited to check the results, detect and confirm patterns, and thereby extend the chart to include lower intermediate scores.

This solution of the problem of points by the area method is similar to one of the solutions of Pascal and Fermat in that each depends on a technique of partitioning. However, instead of partitioning a region, Pascal considered the partitioning of a stake of sixty-four pistoles (units of money). Also, each solution uses a different justification for its partitioning. Of Pascal and Fermat's other two solutions, one relied on combinatorics and the other on the addition of independent probabilities. Secondary school mathematics students could gain some valuable insights into probability by solving the problem of points on their own and then comparing their solution to Pascal and Fermat's three solutions.

Odds of Winning (Bruce:Rita)

	5	6	7	8	9
9	31:1	15:1	7:1	3:1	1:1
8	57:7	26:6	11:5	4:4	1:3
7	99:29	42:22	16:16	5:11	1:7
6	163:93	64:64	22:42	6:26	1:15
5	256:256	93:163	29:99	7:57	1:31

Bruce's intermediate score (vertical axis, rows 9–5)

Rita's intermediate score (horizontal axis, columns 5–9)

Fig. 19.7

An interesting extension to the problem of points occurs if Rita and Bruce use two red marbles and one blue marble in their game. The altered patterns and recursions reflect the influence of the asymmetry induced by the new marble mixture.

An Archery Game

Modeling the analysis on a square provides several pedagogical advantages for solving probability problems. Pictorial representation of the analysis provides visual insights into the concepts of probability. Reliance on geometric skills allows the development of those concepts, which a lack of arithmetic skills would normally impede. Dividing a region in proportion to the appropriate probabilities appeals to students' intuitive understanding of probability. Furthermore, this solution technique provides a mathematical advantage by producing a less complex solution for certain types of sophisticated probability problems. For example, the probability problem presented in the following story involves an infinite Markov chain.

As archers, Rita hits the target 2/5 of the time and Bruce hits the target 1/3 of the time. They decide to have a contest. Letting Bruce shoot first since he is the poorer archer, they alternate shots until one wins by hitting the target. Who is favored? What is each contestant's probability of winning?

Use a square like the one in figure 19.8(a) to calculate the probabilities. Bruce shoots first and has a probability of 1/3 of hitting the target and winning immediately. Color one-third of the square gray, as in figure 19.8(b). If Bruce misses the target, Rita shoots and wins by hitting the target with a probability of 2/5. Of the uncolored region, color two-fifths of it red, as in figure 19.8(c). Notice that the ratio of the area of the gray regions to the area of the red regions is 5:4.

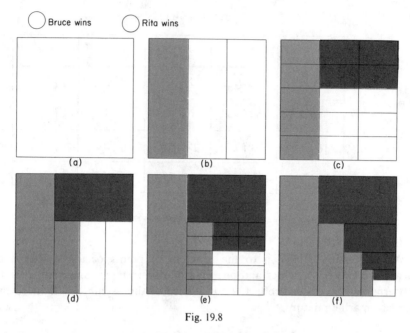

Fig. 19.8

If both shots have missed, Bruce shoots again, and his probability of hitting the target is 1/3. Color one-third of the remaining area gray, as in figure 19.8(d). If no one has hit the target, Rita shoots again, and her probability of hitting the target is 2/5. Color two-fifths of the uncolored region red, as in figure 19.8(e). Notice that for the *newly colored regions* the ratio of the area of the gray regions to that of the red regions is again 5:4. Therefore, for *all the colored regions* the 5:4 ratio is maintained.

Continuing the process, the uncolored region gradually vanishes, and the ratio of the area of the gray regions to the area of the red regions is always 5:4. See figure 19.8(f). Therefore, it is plausible (and correct) to conclude that for this archery contest the ratio of Bruce's chances of winning to Rita's chances of winning is 5:4. Thus, Bruce's probability of winning is

$$\frac{5}{5 + 4}, \text{ or } \frac{5}{9}.$$

Shooting first provides a sufficient advantage to overcome his lesser skill.

The probability problem involved in this archery contest is an example of a Markov chain. The area technique provides a solution that does not require such advanced algebraic processes as matrix multiplication, the summation of infinite series, or the formation and solution of linear equations. Of course, the intuitively appealing conclusion that a particular ratio is maintained throughout an infinite process is assumed but not proved at this level.

20. Geometrical Probability

Richard Dahlke
Robert Fakler

GEOMETRICAL probability is a unique and interesting branch of probability that is replete with applications of high school mathematics. It differs from the discrete probability now taught in some high schools in that it makes use of geometry, plane and analytic, to calculate probabilities.

Why expose students to geometrical probability? To work problems in this subject, students will continually use high school mathematics. Thus, the subject offers them an excellent opportunity to review, reinforce, and apply their previously learned mathematics. The student may have to use the formula for the area of a trapezoid, graph the intersection of linear inequalities, find the relationship among the sides of specific right triangles, or realize that an inscribed angle of a circle is measured by one-half its intercepted arc. Finally, solving some of these problems requires the use of many mathematical concepts, thus giving students the opportunity to see how many isolated concepts contribute to the solution of a significant problem.

When problems in geometrical probability are introduced at appropriate times in a course, students will be able to see direct uses for the mathematics they are learning. This, combined with the interesting nature of the problems themselves, will serve to increase students' interest and enthusiasm.

The Notion of Geometrical Probability

The general problem to be solved can be stated concisely: Given a real-world experiment with outcomes occurring at random and an event, find the probability of a success on a single trial of the experiment. To solve this problem using geometrical probability, experiments must be restricted to those whose outcomes can be represented by points in a geometric region. Then an outcome of the experiment occurring at random corresponds to a point being chosen at random in the geometric region representing the sample space of the experiment. Hence, the original experiment, called a *real-world experiment,* has been transformed into the *mathematical experiment* of randomly choosing a point in a geometric region. Suppose R is the geometric region representing our sample space. Then a given event is represented by a subregion r of R. R is referred to as the sample space of the mathematical experiment and r as an event of the experiment.

▶ **Example 1.** An experiment is conducted by randomly cutting a 5-meter piece of string into two parts. Consider the event that both parts of the string will be at least 1 meter long. Identify the

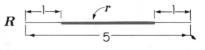

Fig. 20.1

sample space with a line segment of length 5. The corresponding geometrical experiment is choosing a point at random from R. The region R and the subregion r representing the successful outcomes of the experiment are shown in figure 20.1. (Notice that for a successful outcome the string must be cut at a point more than 1 meter from either end of the string. In other words, the randomly chosen point must belong to r.)

▶ **Example 2.** A dartboard consists of a square of side 1 meter that is divided into four squares of side 1/2 meter. Consider the experiment of tossing a dart so that it hits the dartboard at a random spot and the event that the dart hits inside the shaded square shown in figure 20.2. This experiment corresponds to the mathematical experiment of choosing at random a point in a square region R of side 1 unit, and the event corresponds to the chosen point belonging to the shaded square. The region R and the

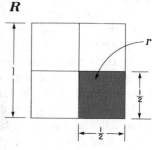

Fig. 20.2

subregion r representing a success are shown in figure 20.2.

Once a real-world experiment has been transformed into a corresponding mathematical experiment with sample space R, the probability of a given event occurring in the real-world experiment will be the probability of the corresponding event r occurring in the mathematical experiment, that is, of a randomly chosen point in R belonging to r. Looking back at examples 1 and 2, what is the probability of the given event occurring in each? In example 1 it seems reasonable that if a random point is chosen in R, the probability p of the point belonging to r is

$$p = \frac{\text{length of } r}{\text{length of } R} = \frac{3}{5}.$$

In example 2 it appears that the probability p of a point randomly chosen inside the larger square being also inside the smaller square is

$$p = \frac{\text{area of } r}{\text{area of } R} = \frac{\left(\frac{1}{2}\right) \cdot \left(\frac{1}{2}\right)}{1 \cdot 1} = \frac{\left(\frac{1}{4}\right)}{1} = \frac{1}{4}.$$

Thus, it can be deduced that for a general mathematical experiment, that is, randomly choosing a point in a region R, the probability that the point chosen will belong to a given subregion r of R is

$$p = \frac{\text{measure of } r}{\text{measure of } R},$$

where measure means length if R is one dimensional and area if R is two dimensional.

Note: Since r is a subregion of R, we have

$$0 \leq \text{ measure of } r \leq \text{ measure of } R.$$

Thus, dividing the members of these inequalities by the measure of R (which is assumed to be positive) gives

$$\frac{0}{\text{measure of } R} \leq \frac{\text{measure of } r}{\text{measure of } R} \leq \frac{\text{measure of } R}{\text{measure of } R},$$

that is, $0 \leq p(r) \leq 1$. This inequality shows that geometrical probabilities are bounded below by 0 and above by 1. Notice that for $p(r) = 0$ the measure of r must be 0 (a point has length 0 and a curve has area 0), and for $p(r) = 1$ the measure of r must equal the measure of R.

▶ **Example 3.** A sample space R and an event r are defined by the diagrams in figure 20.3. Find the probability that a randomly chosen point in R belongs to r.

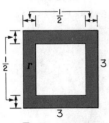

$$\text{Solution: } p(r) = \frac{\text{area of } r}{\text{area of } R} = \frac{3\pi - \pi(1)^2}{3\pi}$$

$$= \frac{2\pi}{3\pi} = \frac{2}{3}.$$

$$\text{Solution: } p(r) = \frac{\text{length of } r}{\text{length of } R}$$

$$= \frac{1 + 2}{6} = \frac{1}{2}.$$

Fig. 20.3

In the diagrams in figure 20.4, a region R and one of its subregions r

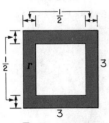

R: Square region
(Ans. $= \frac{5}{9}$)
(a)

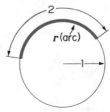

R: Circle
(Ans. $= \frac{1}{\pi}$)
(b)

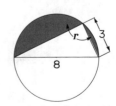

R: Circular region
(Ans. $= (16\pi - 3\sqrt{55})/32\pi$)
(c)

Fig. 20.4

(darkened portion) are represented. See if you can calculate the probability that a randomly chosen point in R also belongs to r.

Direct Applications of Geometrical Probability

▶ **Telephone Line Problem.** A telephone line 50 meters long is suspended between two poles, one containing a transformer, as in figure 20.5(a). Due to a storm, a break occurs at a random point on the telephone line. Find the probability that the break is at a distance not less than 20 meters from the transformer.

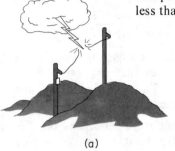

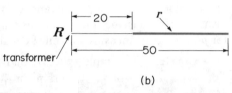

(a)   (b)

Fig. 20.5

Solution. Represent the telephone line by a line segment R of length 50. The line segment r representing the sample space of the experiment and the success region r (those points on the line segment at a distance of not less than 20 from the end representing the pole containing the transformer) are shown in figure 20.5(b). Thus,

$$p(r) = \frac{\text{length of } r}{\text{length of } R} = \frac{30}{50}$$

$$= 0.60 \text{ (or 60 percent)}.$$

▶ **Bomber Problem.** In an attempt to wipe out a munitions area, a bomber is to drop bombs inside a 1-kilometer-square field. At each corner of the field is an abandoned building, as in figure 20.6(a). If a bomb falls within

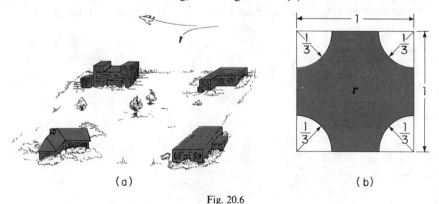

(a)   (b)

Fig. 20.6

1/3 kilometer of any of the buildings, it will be destroyed. Assuming that one bomb is randomly dropped on the field, what is the probability that (a) none of the buildings are destroyed? (b) one building is destroyed? (c) more than one building is destroyed? (d) the bomb falls exactly 1/4 kilometer from a particular building?

Solution. The contact of the bomb tip with the field will be identified with a point, and the field will be represented by a square of side 1.

(a) There is a success (none of the buildings are destroyed) if the distance from the point where the bomb lands to any vertex of the square is at least 1/3 kilometer. The square region representing the sample space of the experiment and the success region r are both shown in figure 20.6(b). The area of the sample space is 1, whereas the area of the success region can be obtained by subtracting the area of the failure region from 1. (The failure region is the complement of the success region. Denote the complement of r by r'.) Now, the area of r' = the area of four quarter circles = $\pi(1/3)^2 = \pi/9$. Thus, the area of the success region is $1 - (\pi/9)$ and

$$p(r) = p(\text{not destroying a building})$$

$$= \frac{\text{area of } r}{\text{area of } R} = \frac{1 - (\pi/9)}{1} = 1 - (\pi/9) = 0.65.$$

(b) A building is destroyed if the bomb falls within 1/3 kilometer of the building. This event is the complement of the event of problem (a); it is the unshaded portion of the diagram. Thus,

$$p(r') = \frac{\text{area of } r'}{\text{area of } R} = \frac{\pi/9}{1} = \pi/9 = 0.35.$$

(c) There is no point of the sample space that is within 1/3 kilometer of at least two buildings. Hence, the event that more than one building is destroyed is an impossible event. The region identified with the event is $\emptyset$, the empty region, and its area is defined to be 0. Thus,

$$p(\emptyset) = \frac{\text{area of } \emptyset}{\text{area of } R} = \frac{0}{1} = 0$$

(d) The success event is a quarter arc of a circle (see fig. 20.7). The area of the arc is 0; thus, the probability is 0. Symbolically,

$$p(r) = p(\text{bomb falls exactly } 1/4 \text{ mile from a building})$$

$$= \frac{\text{area of } r}{\text{area of } R} = \frac{0}{1} = 0.$$

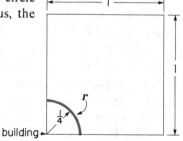

Fig. 20.7

You are invited to work on the following problems. See how many applications of high school mathematics you can find.

1. Consider the Telephone Line Problem.

   (a) What is the probability if 20 is replaced by 40? By 49? By 50? (Answers: 1/5, 1/50, 0.)

   (b) Find the probability if L is substituted for 50 and l for 20.

   (c) By viewing the sample space and successful outcomes, determine what is happening to the probability as l approaches L in value.

   (d) By viewing the formula obtained in (b), determine what is happening to the probability as l approaches L in value. Does this agree with your answer in (c)?

   (e) Graph the probability formula obtained in (b), where p (the probability) is the dependent variable, l is the independent variable, and L is 15. From the graph, analyze the behavior of p as l varies over its domain.

2. For the Bomber Problem, suppose that a building is destroyed if the bomb falls within $1/\sqrt{3}$ km of a building. Find the probability that a randomly dropped bomb will destroy—

   (a) none of the buildings (The challenge in this problem is to find the area of the success region. The failure region consists of four regions of equal area. Each of these regions can be divided into a pair of triangular regions and a sector of a circle (see fig. 20.8). You have enough information to find the area of these regions.) (Answer: $p = 1 - (1/\sqrt{3}) - (\pi/9) = 0.07$)

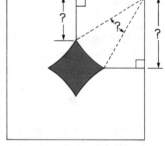

Fig. 20.8

   (b) exactly two buildings (Answer: $(2\pi/9) - 1/\sqrt{3} = 0.12$)

   (c) exactly one building (Answer: $2/\sqrt{3} - \pi/4 = 0.81$)

3. For the Bomber Problem, suppose that a building is destroyed if the bomb falls within x km of a building. What should x be to have a probability of 0.90 that (a) exactly one building is destroyed by a random drop of the bomb on the field? (b) at least one building is destroyed, yet it is possible that no building is destroyed? (c) all the buildings are destroyed? (Answers: $3/\sqrt{10\pi} = 0.54$, $\sqrt{2}/2 = 1.71$, $\geq\sqrt{2}$)

4. For the Bomber Problem, suppose that the field has the shape of an equilateral triangle of side s and that there are three buildings, one at each corner. What is the probability that a bomb randomly dropped on the field will destroy one building if the bomb's destroying radius is $s/3$? (Answer: $2\pi/(9\sqrt{3}) = 0.40$)

Less Direct Applications of Geometrical Probability

It is now time to look at some real-world problems that require a bit more ingenuity in relating the real-world outcome of an experiment to its mathematical counterpart—that of randomly choosing a point in a region. One such problem is a game of chance that you may have played at a carnival or fair and may have wondered about your chances of winning.

▶ **County Fair Problem.** At a county fair a game is played by tossing a coin onto a large table ruled into congruent squares, as in figure 20.9(a). If the coin lands entirely within some square, the player wins a prize. What is the probability that a random toss of the coin will result in a win if *(a)* the coin's diameter is 2 centimeters and the squares have sides of 5 centimeters? *(b)* the coin's diameter is *a* and the squares are of side *b*, where *a* < *b?* (It will be assumed that the markings on the table have no thickness.)

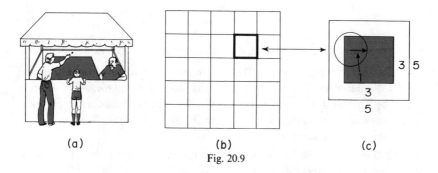

(a)   (b)   (c)

Fig. 20.9

Solution. If possible, translate this problem into one of randomly picking a point from a region. That is, to use geometrical probability the experiment must be equivalent to randomly choosing a point in a region. This is accomplished by focusing our attention on where the coin's *center* lies on the table. Thus, *each outcome of the experiment will be the position of the coin's center on or within whatever square it falls.* Such a square is shown in figure 20.9(b). Since this square and its interior is the sample space, it is isolated in figure 20.9(c).

   (a) A subregion containing the successful outcomes must be found. To have a success (that is, the coin is interior to the square), the center of the coin must lie at least the radius length (that is, 1 centimeter) from the boundary of the square region. The success region is the square region in figure 20.9(c) of side 5 − (1 + 1) = 3. Therefore,

$$p = \frac{(3)^2}{5^2} = \frac{9}{25}.$$

(b) For a success, the center of the coin must lie at least $a/2$ units (the radius length) from the boundary of the square. The success region is the square region in figure 20.10 of side $b - (a/2 + a/2) = b - a$. Therefore,

$$p = \frac{(b - a)^2}{b^2}.$$

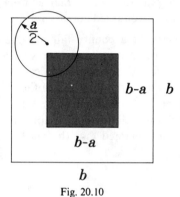

Fig. 20.10

Here are some exercises based on the County Fair Problem.

1. Discuss, by viewing figure 20.10, how the probability varies as the coin's diameter approaches in length the side of a square. Also, discuss how the probability varies as the coin's diameter approaches 0. Support your answers by analyzing the formula $p = (b - a)^2/b^2$ as a approaches b and as a approaches 0.

2. Graph the formula in *(a)* where b is constant, a is the independent variable, and p is the dependent variable. *(Note:* $0 < a < b$)

3. If $b = 1/2$ cm, what size coin should a customer throw so that the probability of winning a prize is 0.36? (Answer: $a = 0.20$ cm)

4. If $a = 1/4$ cm, what should a side of a square be so that the probability of a failure is 0.19? (Answer: $b = 2.50$ cm)

The Coordinate System in Geometrical Probability

The coordinate system, which plays a major role in high school mathematics, is often essential for solving problems in geometrical probability. More specifically, the coordinate system (one or two dimensional) is needed to construct a representation of a sample space and success event for an experiment.

▶ **Triangle Problem.** A line segment AB, with midpoint M, has length a. A point X is randomly chosen interior to the segment. What is the probability that $\overline{AX}$, $\overline{BX}$, and $\overline{AM}$ can be the sides of a triangle?

Solution. Coordinatize segment AB where A is labeled with 0, B with a, and X with x. Thus, M has coordinate $a/2$; see figure 20.11(a). The sample space consists of all points with coordinate x, where $\overline{AX}$, $\overline{BX}$, and $\overline{AM}$ can be the sides of a triangle. What is a necessary and sufficient condition for these segments to be the sides of a triangle? The answer lies in a plane geometry result, namely, the sum of the lengths of any two sides must be greater than the third side. Since the length of $\overline{AX} = x$, the length of $\overline{BX} = a - x$, and the length of $\overline{AM} = a/2$,

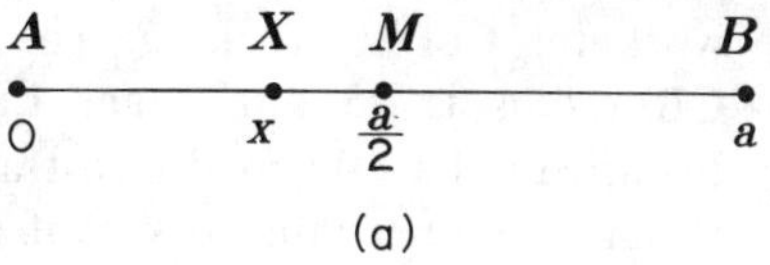

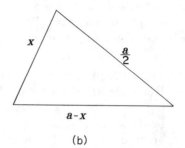

(a)

(b)

Fig. 20.11

$$x + (a - x) > \frac{a}{2}$$

$$x + \frac{a}{2} > a - x$$

$$(a - x) + \frac{a}{2} > x.$$

See figure 20.11(b). This system of linear inequalities is equivalent to the system

$$a > \frac{a}{2} \text{ (which is always true)}$$

$$x > \frac{a}{4}$$

$$x < \frac{3a}{4}.$$

Writing this another way, a success occurs if the point selected has coordinate x, where $a/4 < x < 3a/4$. The sample space R and success region r are shown in figure 20.12. We get

$$p = \frac{\dfrac{3a}{4} - \dfrac{a}{4}}{a} = \frac{\dfrac{a}{2}}{a} = \frac{1}{2}.$$

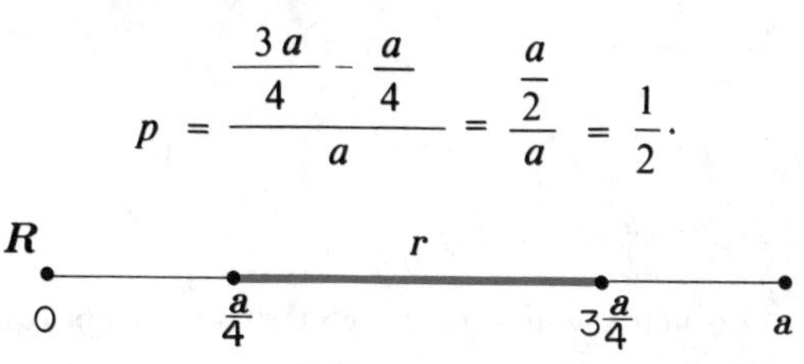

R r

0     $\frac{a}{4}$ $3\frac{a}{4}$ a

Fig. 20.12

▶ **CB Radio Problem.** Two CB (Citizen Band) radio operators, Tiger Lily and Huge Huey, work for Carl's Trucking. The range of their CB radios is 25 km. Tiger Lily is traveling toward the base from the east and at 3:00 P.M. is somewhere within 30 km of the base. Huge Huey is traveling toward the base from the north and at 3:00 P.M. is somewhere within 40 km of the base. What is the probability that they can communicate with each other at 3:00 P.M.?

Solution. Let x and y represent the distances that Tiger Lily and Huge Huey are from the truck base, respectively. Hence, $0 \le x \le 30$ and $0 \le y \le 40$. Now, the set of all pairs of distances can be represented by the set of ordered pairs (x,y), with the given restrictions on x and y. The graph of this set of ordered pairs is our sample space. Each point of the graph represents a particular combination of positions for Tiger Lily and Huge Huey. They can communicate over their radios if the distance between them is not more than 25 km; see figure 20.13(a). Thus, the points of the sample space that constitute a success are represented by the ordered pairs (x,y) satisfying the inequality $\sqrt{x^2 + y^2} < 25$, which is equivalent to $x^2 + y^2 < 625$. The region representing the sample space and the shaded success region are shown by figure 20.13(b). The area of the region representing the sample space is 1200 km², and the area of the success region is $(1/4)\pi(25)^2 = 625\pi/4$. Thus,

$$p = \frac{\dfrac{625\pi}{4}}{1200} = \frac{625\pi}{4800} = 0.41.$$

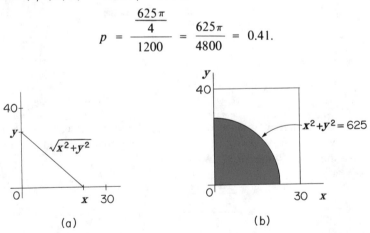

Fig. 20.13

The following problems will require you to use a coordinate system to represent the sample space and success event.

1. A line segment AB of length a has a fixed point C belonging to its interior.

    (a) If a point X is randomly chosen interior to $\overline{AB}$, what is the probability that $\overline{AX}$, $\overline{XB}$, and $\overline{AC}$ can form a triangle? An equilateral triangle? (*Hint:* Let b (a constant) and x be the coordinates of C and X, respectively.) (Answers: b/a, 0)

    (b) Use the probability you obtained as the answer to (a) to verify the answer obtained in the first example of this section, where C is the midpoint of $\overline{AB}$.

2. A point X is chosen at random from the interior of a line segment of length 1 cm. What is the probability that the product of the lengths of the two segments formed is less than 3/16? Less than 1/4? Less than 1/8? Greater than 1/16? (Answers: $1/2 = 0.5$, 1, $(2 - \sqrt{2})/2 = 0.29, \sqrt{3}/2 = 0.87$)

3. A bus of line A arrives at a station every four minutes and a bus of line B every six minutes. A potential passenger of line A just arrived at the station and is hoping that the next bus back to the station is from line A. What is the probability that the next bus arriving is from line A? That a bus of any line arrives within two minutes? (*Hint:* Let x and y be the number of minutes the passenger has to wait for a bus of line A and a bus of line B, respectively.) (Answers: $2/3 = 0.67$, 2/3)

4. José and Juanita Juarez have a joint checking account in the amount of \$200. One day, unknown to each other, they each write checks on this account for random amounts not to exceed \$200. Find the probability that—

    (a) more than \$50 remains in the account (Answer: $9/32 \approx 0.28$)

    (b) their account is overdrawn (Answer: $1/2 = 0.5$)

    (c) their account is not overdrawn by more than \$200 (Answer: 1)

5. Two rival automobile manufacturers agree to have a public competition in which both of them will try to prove that their own compact car (Car A or Car B) gets the better gas mileage. The winning car is to be the one going farther on 5 gallons of gas. On the night before the competition both drivers, in order to win, secretly siphon a random amount of gas (up to one gallon) out of their opponent's car. What is the probability that Car A wins if Car A gets 25 miles per gallon and Car B gets 30 miles per gallon? (Answer: $1/60 = 0.02$)

21. Evaluating Exact Binomial and Poisson Probabilities without Tables

Harry O. Posten

IN THE usual introductory probability or statistics course, the basic discrete distributions are introduced through, at most, a limited set of tables. Even for the binomial distribution, the general rule is to provide tables for a few values of N and restrict problems to this set or to use a normal approximation. This approach is somewhat unsatisfactory, since it fails to show students how to handle future binomial problems except by means of the approximation. Most students have considerable difficulty with the normal approximation unless a large amount of time is devoted to this topic— and even then the issue is in doubt. Additionally, whatever the teacher's feeling concerning the use of limited tables and the normal approximation, this approach is usually lacking in excitement.

Fortunately, there is a supplementary or alternative approach that permits the use of more exciting "real" data to introduce the student to one or more of the basic discrete distributions. This approach is available to any teacher with access to any kind of computer, including a programmable hand calculator. The teacher need have no programming experience, since the algorithm is simple and can be programmed by a colleague or student with limited knowledge of computer programming. (Its simplicity, though, might conceivably encourage a teacher to pick up some programming skills.) The same basic algorithm is valid for the binomial and Poisson distribution as well as for several other important discrete distributions *(hypergeometric, geometric,* and *negative binomial).* This article will present the algorithm and its use and will include two worksheets that can be used to stimulate students in these topics. The worksheets can be photocopied and ditto masters made directly from them without retyping.

Evaluating Binomial and Poisson Probabilities Using a Simple Recursive Algorithm

The two most important discrete distributions are the binomial and Poisson distributions. Both can be evaluated by a simple, easily understood recursive algorithm. This algorithm can also be used to evaluate probabilities for the three other discrete distributions mentioned above; however, extensions to these distributions will be left to the interested reader. For the binomial and Poisson distributions, the mathematical forms are

154

binomial	**Poisson**	
$f(x) = \dfrac{N!}{x!\,(N-x)!}\,p^x q^{N-x}$	$f(x) = \dfrac{e^{-\lambda}\,\lambda^x}{x!}$,	(1)

$(p$ = the probability of an event
occurring, and $p + q = 1)$

with the domain of each function being the nonnegative integers and, in the binomial case, with x restricted above by the value $x = N$. For each of these distributions, it is obvious that $f(0)$ is easily evaluated on any computer, since $f(0)$ is q^N and $e^{-\lambda}$, respectively. The basic property of each distribution that permits a simple recursive evaluation of probabilities is the fact that the probability for any value of x is a simple multiplier times the probability for the previous value $x - 1$. The reader may readily verify the following:

binomial	**Poisson**	
$f(x+1) = \dfrac{p(N-x)}{q(x+1)}\,f(x)$	$f(x+1) = \dfrac{\lambda}{x+1}\,f(x)$	(2)
$x = 1, 2, \ldots, N$	$x = 1, 2, \ldots$	

Consequently, if we let $c(x)$ be defined by

binomial	**Poisson**	
$c(x) = \dfrac{p(N-x)}{q(x+1)}$	$c(x) = \dfrac{\lambda}{x+1}$,	(3)

we may use the recursive procedure of evaluating $f(0)$ directly and then evaluating $f(k) = c(k-1)f(k-1)$ for values of k equal to 1, 2, and so forth, until $k = x$ to produce the value of $f(x)$. Because the computations involved are relatively trivial, the algorithm is quite fast and may be used to evaluate cumulative probabilities as well as individual probabilities by accumulating the latter at each recursive step. A good recommendation is to use this algorithm to provide a single program to produce as its output the individual probability and cumulative probability for arbitrary values of x. For computers with printed output this algorithm is most convenient for producing a table by printing out the value of x and the individual and cumulative probabilities at each step. The algorithm for both the binomial and the Poisson distributions is given in table 21.1 with $c(x)$ defined by (3).

To produce a binomial or Poisson table, the algorithm in table 21.1 needs only to be modified to allow the output of $k, f(k), F(k)$ for $k =$

TABLE 21.1

ALGORITHM FOR BINOMIAL AND POISSON PROBABILITIES

Step 1:	Enter value of x and the distribution parameters (N,p for binomial and λ for Poisson).
Step 2:	Evaluate $f(0)$.
Step 3:	Recursively evaluate $f(k) = c(k-1)f(k-1)$ and the cumulative probability $F(k) = F(k-1) + f(k)$ for $k = 1, 2, \ldots, x$.
Step 4:	Print x, $f(x)$, $F(x)$.

0, 1, 2, ... at step 2 and each stage of step 3. Additionally, x should not be entered in step 1, since all probabilities are being tabulated. For the Poisson distribution, of course, the program should include a statement that will force the calculations to terminate when essentially all the probability has been accumulated.

Applications

The teacher will make most effective use of the foregoing algorithm as a teaching tool in an environment where a program can be stored by the teacher and easily executed by a student. The teacher may then assign any kind of realistic problem without concern for the availability of appropriate tables or the problem of approximation. The focus can be entirely on the problem itself without extraneous distractions. In particular, the teacher may provide sets of real data (with observed relative frequencies already calculated) and ask the student to generate the proper table of probabilities for comparison. The closeness of the agreement between the probabilities and the observed relative frequencies indicates how closely the data follow the probability law. This is a most effective way to introduce a distribution and to provide students with a hands-on feel for a given probability law while at the same time showing them that a mathematical model can accurately describe the random pattern of real data. It is necessary, of course, for the teacher to provide proper data from the literature and other sources. A particularly good source of data from the biological area is the book *Biometry, the Principles and Practice of Statistics in Biological Research* [126]. Data also are provided in the two worksheets in this article.

Worksheet 1 applies the binomial probability law to the game of baseball. Many students find this problem quite interesting because they can readily relate to it. In fact, this problem was considered of sufficient general interest in 1961 to justify a feature article in *Life* magazine. The cover pictured two baseball players, Mickey Mantle and Roger Maris, both of whom were attempting that year to break Babe Ruth's long-standing record of sixty home runs in a single season. Inscribed above another picture of the players was the formula for the binomial distribution.

Worksheet 2 consists of three good sets of classical data, each of which closely follows the Poisson law. For each data set the observed frequencies for the various values of the random variable ("counts") are provided along with the observed relative frequencies. The mean of the data is also provided, and the teacher only needs to assign the student the problem of filling in, for comparison, the probabilities for a Poisson random variable having a mean identical with the observed mean (i.e., λ = observed mean). These probabilities are obtained by generating the required Poisson table using the previous algorithm modified to produce a table. The author has found students particularly impressed with this teaching strategy because they actually work with a Poisson distribution and with real data following this law.

The problems on the worksheet are easily solved without recourse to tables or approximation by executing a binomial program based on the foregoing algorithm, though it is sometimes pedagogically useful to solve the problems approximately as well to show the student the quality of the normal approximation. This example is particularly useful in encouraging students to think about the conditions behind the binomial probability law and to understand more clearly the model involved. Often they will argue with the teacher about the validity of the assumptions of independence and of constant probability of success. The author's experience in baseball suggests that these assumptions are reasonably rough approximations, particularly for the Mantle and Maris environment, since the question being asked is "What are the probabilities if the process is stable and the players continue to hit at their present rate?" In any event, this is a natural problem to raise questions concerning underlying assumptions and to reinforce the initial presentation of the binomial probability model. The existence of a convenient program to calculate the probabilities allows the teacher and students to concentrate on the main issues.

A warning and a modification

On reflection, the algorithm presented in this article is seen to have one obvious flaw. Should $f(0)$ have a value smaller than the smallest magnitude allowed by the computer, underflow will occur and $f(0)$ will be read as zero by the computer. In this event, all subsequent probabilities, $f(k)$, will be calculated as zero, since each of these probabilities is calculated as a multiplier times the previously calculated probability. Such an error can be avoided by incorporating into the program at step 1 of the algorithm a statement that stops the program when $f(0)$ is read as zero and informs the user to use another method. Since for the binomial distribution $f(0) = q^N$ may be sufficiently small when q is very small with N large, this is a practical consideration. Also, for the Poisson distribution, $f(0) = e^{-\lambda}$ may be sufficiently small when the mean λ is large. Consequently, it is wise to provide a program termination when such a problem occurs and then use a backup program. An effective backup procedure for such cases is a modification of the foregoing algorithm which uses a natural logarithm and an exponential operation at each step. Specifically, in step 1 ln $f(0)$ is evaluated instead of $f(0)$, and in step 3 the natural logarithm of $f(k)$ is evaluated by ln $f(k)$ = ln $c(k-1)$ + ln $f(k-1)$. The procedure is now recursive in the logarithm of $f(k)$, and recursive addition avoids the difficulty found in recursive multiplication. At each stage, ln $f(k)$ must be converted to $f(k)$ by the exponentiation $f(k)$ = exp [ln $f(k)$] for $k = 0, 1, \ldots,$ x to evaluate the cumulative probability $F(x)$. This modified algorithm is given in table 21.2 and is easily transformed to provide a table of probabilities in the same manner as before. When algorithm 1 is invalid because of $f(0)$ being calculated as zero, the algorithm of table 21.2 will be an effective alternative.

TABLE 21.2

MODIFIED ALGORITHM FOR BINOMIAL AND POISSON PROBABILITIES

Step 1:	Enter value of x and the distribution parameters (N,p for binomial and λ for Poisson).
Step 2:	Evaluate ln $f(0)$ (= N ln q for binomial and $-\lambda$ for Poisson) and $f(0)$ = exp [ln $f(0)$].
Step 3:	Evaluate ln $f(k)$ = ln $c(k-1)$ + ln $f(k-1)$, $f(k)$ = exp [ln $f(k)$], and $F(k)$ = $F(k-1)$ + $f(k)$ for k = 1, 2, ..., x.
Step 4:	Print $x, f(x), F(x)$.

Algorithms 1 and 2 are easy to explain, and student understanding may be enhanced by the assignment of pedagogically useful names—for example, BINRC, BINRT, BINLRC, and BINLRT, where BIN refers to binomial, R to recursive, LR to log-recursive, and C and T to cumulative and table.

Extensions to other discrete probability distributions

The algorithms of tables 21.1 and 21.2, with a proper choice for $c(x)$, are actually valid for any discrete probability distribution having the property that the probability for any value of the random variable can be expressed as a simple multiplying factor times the previous value of the random variable. This includes the geometric, negative binomial, and hypergeometric distributions. The programs for the first two are easy to write, but the hypergeometric algorithm has some inherent difficulties because one must specify a convenient method for evaluating the initial value (e.g., $f(0)$) and must take into account that the range of the random variable is restricted in some cases.

WORKSHEET 1

Binomial Probability Problem
Maris-Mantle Example

During the 1961 baseball season, two players on the same American League team appeared to have a chance (after two-thirds of the season was completed) of breaking the old record of sixty home runs in a single season held by Babe Ruth for many years. These two players were Mickey Mantle and Roger Maris of the New York Yankees. Their records at the end of 106 games are given below:

Player	Home Runs (HR)	At Bats (AB)	HR/AB
Mantle	40	387	.1034
Maris	41	414	.0990

Babe Ruth made his record when teams played 154 games a season. The 1961 season was the first in which the teams played an extended schedule of 162 games. Consequently it was of interest to know what the chances were that either player would break the record in 154 games and what the chances were for 162 games. On the basis of the first 106 games, one may calculate the approximate number of at bats each player has left by determining the average number of at bats per game for each player. The result is 3.65 AB/game for Mantle and 3.91 for Maris. This leads to the following table of approximate at bats left, assuming each player plays all the remaining games.

Number of AB Left

Player	154 Games	162 Games
Mantle	175	204
Maris	188	219

Since the number of at bats left amounts to a series of N trials, each of which has only two possible outcomes—home run or no home run—if we assume independence and a constant chance of a given player hitting a home run on each trial, what we are concerned with is a binomial random variable for each player; X = number of home runs in N trials. On this basis, calculate the following:

1. The probability that Mantle will *break* the record
2. The probability that Maris will *break* the record
3. The probability that neither will break the record
4. The probability that at least one will break the record
5. The probability that both will break the record

Calculate all probabilities for 154- and 162-game schedules.

Answers

Players	154 Games	162 Games
Mantle		
Maris		
Neither		
1 or 2		
Both		

WORKSHEET 2

Examples of Random Data That Follow
a Poisson Probability Law

1. In the famous Rutherford and Geiger (1930) experiment quoted by Fisz [41, p. 143], Rutherford and Geiger observed the number of α-particles emitted by a radioactive substance in $N = 2608$ periods of 7.5 seconds each.

X = Number of Particles	Observed Number of Periods	Observed Proportion	Theoretical Proportions
0	57	.0219	
1	203	.0778	
2	383	.1469	
3	525	.2013	
4	532	.2040	
5	408	.1564	
6	273	.1047	
7	139	.0533	
8	45	.0173	
9	27	.0104	
10	16	.0061	
Total	2608		

The average number of particles per time period in this experiment is $\lambda = 3.87$; so one may obtain theoretical proportions (probabilities) from the Poisson probability law $P(X = x) = e^{-\lambda}\lambda^x/x!$ using this value of λ.

2. Bortkiewicz *(Das Gesetz der Kleinen Zahlen,* Leipzig: Teubner, 1898), quoted by Fisz [41, p. 141], considered Prussian army data on horse-kick deaths in a one-year period in 10 army corps over twenty years (a total, in effect, of 200 corps).

X = Number of Deaths	Number of Corps	Observed Proportion	Theoretical Proportions
0	109	0.5450	
1	65	0.3250	
2	22	0.1100	
3	3	0.0150	
4	1	0.0050	
Total	200		

The average number of horse-kick deaths per army corps per year in this data is $\lambda = 0.61$; so one may obtain theoretical proportions (probabilities) from the Poisson probability law using this value of λ.

(Continued on next page)

WORKSHEET 2—Continued

3. Clarke [16] gave the number of flying bombs falling into 1/4-square kilometer areas of London over a certain period of World War II. There were 576 such areas.

X = Number of Hits	Number of Sectors	Observed Proportion	Theoretical Proportions
0	229	.3976	
1	211	.3663	
2	93	.1615	
3	35	.0608	
4	7	.0122	
5	0	.0000	
6	0	.0000	
7	1	.0017	
8	0	.0000	
Total	576		

The average number of hits per sector (per 1/4 km^2) for the data is $\lambda = .932$; so one can obtain theoretical proportions (probabilities) from the Poisson probability law using this value of λ.

V. Applications

22. Paradoxes in Sampling

Clifford H. Wagner

MANY paradoxes and questions in probability theory can be resolved by the correct use of conditional probability. Here are two examples of sampling paradoxes that illustrate the use of conditional probabilities, reveal glimpses of probability and statistics in the early twentieth century, and relate to certain contemporary issues of justice and public opinion surveys.

Keynes's Principle of Indifference

John Maynard Keynes [71] was a student and philosopher of probability as well as a famous economist. The principle of indifference is the name Keynes gave in 1921 to the strategy of assigning equal probabilities to all outcomes when we have no reason to differentiate among the likelihoods of the various outcomes. Jakob Bernoulli, a founder of probability theory, first enunciated this strategy, which is also known as the principle of non-sufficient reason and as the principle of equiprobable outcomes. By whatever name it is known, the principle of indifference is and always has been an essential tool of probability; however, the misapplication of this principle frequently causes paradoxical and contradictory conclusions.

Consider the following game invented and played by two hypothetical but inveterate gamblers. The first player (whom we shall call the holder) fills a fishbowl with blue and yellow marbles in equal proportions and holds the bowl while the second player (the drawer) closes her eyes and randomly selects a sample of two marbles. If the drawer selects two marbles of the same color, she wins five dollars; if the marbles do not match, the holder wins five dollars. The two gamblers who invented this game still play it with zest and determination, the same player always holding

162

the bowl and the other always drawing the marbles, because each player believes that in this way he or she will have at least a fair chance of winning.

The drawer has reasoned as follows: "There are three outcomes—either both marbles are blue, both are yellow, or the colors are different. By the principle of indifference, each outcome has probability 1/3, and since two outcomes favor me, the probability that I will win is 2/3. I will keep playing until my opponent realizes his disadvantage."

The holder (who is less greedy) has reasoned: "Since the marbles are selected without replacement, there are really four outcomes—both marbles are blue, both are yellow, the first is blue and the second yellow, or the first is yellow and the second blue. Because the principle of indifference implies that each outcome has probability 1/4 and two outcomes favor me, the probability that I will win is 2/4. This is a fair game and I will keep playing because I love to gamble."

Although the holder expects to break even in the long run and the drawer expects a net profit, both players have reasoned incorrectly. In resolving the paradox of their contradictory conclusions, we shall find that the probability of matching colors, and thus also the probability that the drawer wins, is actually less than 1/2. (Despite the famous opinion of Leo Durocher, it is possible and sometimes even probable that nice guys finish first.) The error of our players is typical of many inexperienced and even some experienced students of probability—they did not properly enumerate the set of all possible outcomes.

There are several ways of finding the correct probability of a match, and they all involve specifying the total number of marbles, which we shall represent by n. As it turns out, for very large values of n the game is almost fair, but for small values of n the holder has a significant advantage. To calculate an answer using the principle of indifference, we must be able to consider the sample space as a set of equally likely outcomes. For example, imagine that distinguishing numbers from 1 to $n/2$ have been marked on the blue marbles and again on the yellow marbles. Then there are $\binom{n}{2}$ possible pairs to be selected, each outcome having probability $1/\binom{n}{2} = 2/n(n-1)$. The number of all-blue pairs is $\binom{n/2}{2}$ and so is the number of all-yellow pairs; therefore the probability of a match is

$$P(\text{Match}) = 2 \cdot \binom{n/2}{2} \cdot \frac{2}{n(n-1)} = \frac{n-2}{2n-2} .$$

The same result can be obtained more simply using conditional probability. This second method avoids the risky principle of indifference and is easily generalized to other problems. It is convenient to restate the definition of conditional probability in the form of a theorem, the multiplication rule for two events, X and Y:

$$P(X \text{ and } Y) = P(X \cap Y) = P(X)P(Y \mid X).$$

Now, the probability of selecting two blue marbles (B) is the product of the probability that the first marble is blue and the conditional probability that the second marble is blue given that the first is blue, that is,

$$P(B_1 \text{ and } B_2) = P(B_1)P(B_2 \mid B_1)$$

$$= \frac{n/2}{n} \cdot \frac{n/2 - 1}{n - 1} = \frac{1}{2} \cdot \frac{n - 2}{2n - 2}.$$

The probability of drawing two yellow marbles (Y) is exactly the same, and thus

$$P(\text{Match}) = P(B_1 \text{ and } B_2) + P(Y_1 \text{ and } Y_2) = \frac{n - 2}{2n - 2}.$$

It is evident from table 22.1 that the probability of a match appears to be always less than $1/2$ but approaches $1/2$ when n is large. It is not difficult to prove that the inequality

$$\frac{n - 2}{2n - 2} < \frac{1}{2}$$

holds whenever $n > 1$ and that also $\lim\limits_{n \to \infty} P(\text{Match}) = 1/2$.

TABLE 22.1

THE PROBABILITY OF MATCHING COLORS

Number of Marbles, n	$P(\text{Match})$, $\dfrac{n - 2}{(2n - 2)}$
4	.33
6	.40
8	.43
10	.44
20	.47
50	.49
100	.495
1 000	.499 5
10 000	.499 95

The "Literary Digest" Poll

Our second example involves the problem of finding the correct explanation for a spectacular mistake. Darrell Huff, George Gallup, and many authors of statistics textbooks have reported the example of the disastrous *Literary Digest* poll of 1936, which predicted that Alfred Landon would defeat Franklin Roosevelt by 54 percent to 41 percent (Roosevelt won a landslide victory and actually received 61 percent of the popular vote). Several commentators have suggested that the poll's error, and perhaps even the subsequent demise of the magazine, was caused by a sample selec-

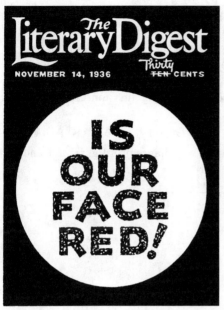

tion bias in which the opinions of automobile owners and telephone subscribers were overrepresented and the opinions of the economically disadvantaged were underrepresented.

More recently, Maurice Bryson [11] has called the selection bias explanation a "statistical myth"and has suggested that the magnitude of the error (which was 61 – 41 = 20 percentage points) can only be explained by a nonresponse bias resulting from the return of only 2.4 million ballots (approximately) out of 10 million ballots distributed by the *Literary Digest.*

What is the explanation for such a large contradiction between the predictions, based on the magazine's sample, and the observed outcome of the election? A postmortem report in the *Literary Digest* attempted to refute the obvious criticism of selection bias. As that report states,

> The "have-nots" did not reelect Mr. Roosevelt. . . . The fact remains that a majority of farmers, doctors, grocers, and candlestick-makers *also* voted for the President. As Dorothy Thompson remarked in the New York *Herald Tribune,* you could eliminate the straight labor vote, the relief vote and the Negro vote, and *still* Mr. Roosevelt would have a majority.

Moreover, the *Digest* did not rely just on auto registrations and telephone lists for obtaining their sample. In Chicago, they polled one-third of all registered voters, and in some cities, such as Allentown, Pennsylvania, they polled every registered voter. In every city, the returned ballots indicated a Landon victory, whereas the actual election results in each case supported Roosevelt. The error in estimating Roosevelt's percentage was 18 points in Chicago and 12 points in Allentown. The *Digest*'s selection procedure was woefully biased, but the very large errors that occurred, even in cities where the selection method was more thorough, suggest that selection bias alone does not account for the large overall error of 20 percentage points.

To demonstrate the significance of the nonresponse bias among the individuals selected for the poll, let us first assume that there was no selection bias in the sampling procedure and compare the preferences of those who responded with the preferences of those who did not. Accordingly, we can define the following two events for the sample space comprising all registered voters who received ballots for the poll:

R = a voter who is a Roosevelt supporter
A = a voter who is willing to answer the *Digest* poll

From the final election results, let us use $P(R) \approx .61$, and from the poll results we have $P(A) \approx$ 2.4 million/10 million $= .24$, and the conditional probability $P(R|A) = .41$.

To estimate $P(R|A')$, Roosevelt's support among the nonrespondents, we need the theorem

$$P(R) = P(R \cap A) + P(R \cap A') = P(A)P(R|A) + P(A')P(R|A'). \quad (1)$$

The proof of this theorem is based on the Venn diagram in figure 22.1 and the multiplication rule.

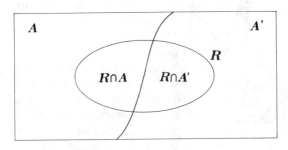

Fig. 22.1. A partition for the set *R*

Solving (1) for $P(R|A')$ leads to

$$P(R|A') = \frac{P(R) - P(A)P(R|A)}{P(A')} = \frac{.61 - (.24)(.41)}{.76} \approx .67.$$

That is, although only 41 percent of those who answered the poll were Roosevelt supporters, approximately 67 percent of the nonrespondents (who were a majority of those contacted) supported Roosevelt. The bottom row of entries in table 22.2 was obtained by similar calculations and indicates a sharp contrast between the political attitudes of those who answered and those who did not.

TABLE 22.2
A COMPARISON OF RESPONDENTS AND NONRESPONDENTS

	Percent Support for Candidate		
	Roosevelt	Landon	Others
Among Respondents, A	41	54	5
Among Nonrespondents, A'	67	31	2

One explanation for the nonresponse bias would be that the supporters of Landon or minor party nominees sensed the weakness of their candidate and felt an urgent need to show their support, whereas Roosevelt supporters may have been more confident and complacent. We can find evidence for this hypothesis by calculating another sequence of conditional probabilities. For example, to estimate the proportion of Roosevelt supporters who chose to answer the poll, we can use the definition of conditional probability and the multiplication rule to find

$$P(A \mid R) = \frac{P(A \cap R)}{P(R)} = \frac{P(A)P(R \mid A)}{P(R)} = \frac{(.24)(.41)}{(.61)} \approx .16.$$

Apparently, only 16 percent of the Roosevelt supporters who were contacted were willing to answer, whereas similar calculations show that 35 percent of the Landon supporters and 40 percent of the supporters of the minor candidates were willing to answer the poll.

All these calculations should be considered rough estimates because the existence of selection bias in the choosing of voters to be polled means that we cannot really know $P(R)$, the proportion of Roosevelt supporters among the 10 million voters contacted. Our calculations were based on an approximation, $P(R) \approx .61$. But even if we had used a value as small as .43, we would still have found evidence to support the two trends: Roosevelt had greater support among the nonrespondents, and Roosevelt supporters were less willing to answer the poll.

Present-Day Connections

The gamblers of the first example each ignored how the color of the second marble (that is, whether it matches or not) depends on the color of the first marble. A similar error occurred in a widely publicized trial in California where a couple was convicted in a mugging case on the circumstantial evidence that they possessed six characteristics reported by witnesses. The prosecutor had argued that the odds of any couple having these six characteristics were 1 in 12 million; hence, this couple was the guilty couple beyond a reasonable doubt. The prosecutor's calculation of probabilities was done by estimating a probability for each distinct characteristic, and then multiplying the six probabilities as if the characteristics were mutually independent. This conviction, described as a "Trial by Mathematics" in *Time* magazine [134], was reversed by the California Supreme Court, in part because the prosecutor had incorrectly used the multiplication rule.

Since the days of the *Literary Digest*, professional pollsters have developed sophisticated sampling techniques that yield very reliable estimates. Nevertheless, simple mail-back survey questionnaires are still used by newspapers, legislators, and various organizations. Both the publishers of such questionnaires and their readers or constituents must recognize the inherent nonresponse bias and should never interpret the responses to such surveys as accurate representations of the entire population.

23. Using "Consumer Reports" as a Resource for Data Analysis in the Statistics Classroom

John W. McConnell

A MAGNIFICENT resource is available to statistics teachers in their school libraries. *Consumer Reports* [26], a monthly magazine published by Consumers Union (CU), provides evaluations of commonly used products. A teacher can find many articles in this magazine that rate cameras, foods, automobiles, stereos, TVs, skis, tennis rackets, and other items of interest to teenaged consumers. The articles provide the data that students at all levels of statistical sophistication can use to test hypotheses about products. In addition, the reports detail the procedures used by CU to produce their evaluations of brand-name products. These procedures illuminate the problem of quantifying quality. They add to the students' understanding of product evaluation and promote a healthy, investigative approach to consumerism.

The two kinds of product evaluation presented here have been used with high school students. The first example focuses on the use of data to investigate a question—specifically, are more expensive products better products? The second example illustrates the issues students can consider as they review CU's techniques for making human ratings less subjective.

Quality and Price

Peter C. Riesz, a marketing professor at the University of Iowa, considered the hypothesis "You get what you pay for" through an exhaustive examination of *Consumer Reports* for 1961 to 1975. Riesz related product costs to product ratings for 685 categories of consumer products. Ratings and costs can be tracked down through the *Consumer Reports Buying Guide Issue,* which each December summarizes the product ratings from the previous two years. We test, for example, the relationship between the list price and the accuracy for medium-priced loudspeakers using data from the 1979 *Buying Guide* (see table 23.1).

The data provide students with an opportunity to investigate the hypothesis with a graph. The hypothesis implies that accuracy depends on price. Therefore, a graph should use accuracy as the dependent variable *(y)* and cost as the independent variable *(x)*. The graph (an exercise for

TABLE 23.1
ACCURACIES AND LIST PRICES OF
FIFTEEN MEDIUM-PRICED LOUDSPEAKERS

Brand	Accuracy	List Price
AVID 102	91	$130
EPI 100	90	100
BIC Venturi	88	120
Marantz Imperial	87	100
Lafayette Criterion	85	100
Micro-Acoustics	85	129
Realistic Optimus	84	115
Dynaco	84	130
Tempest	84	131
Janszen	83	120
KLH	81	130
Altec	80	119
Audioanalyst	78	107
Advent	78	120
Rectilinear	67	109

the reader) fails to show a relationship between the two variables, particularly because of the high accuracy ratings of the three least expensive speakers. I use the data with juniors and seniors in a statistics course that introduces correlation and regression early. These students then compute the correlation ($r = .08$), which confirms the graphical results—no relationship between quality and price.

Students challenge this result. "These are list prices. Nobody pays list. How much are these speakers when discounted?" "These are medium-priced loudspeakers; so the list price doesn't vary much. What if we included some very expensive and some very cheap speakers?" These questions provide some plausible rival hypotheses for the conclusion reached by examining the graph. They are good questions and can be tested. I ask some students to visit local audio stores to get recent prices so that our data represent current discounted prices. We investigate the second question using the 1979 *Buying Guide* (pp. 176–83) as a resource. Ratings of fifteen expensive and fourteen inexpensive speakers are available there. The new data broaden the ranges of the price and accuracy variables. When the total information on the forty-four speakers is graphed as ordered pairs, a relationship between quality and price does exist ($r = .49$). This exercise can be adapted for students not ready for the mathematics of correlation and regression. Even general mathematics students can compute the average of the ratings assigned to each of the three groups (expensive, medium, inexpensive) to show that a person willing to spend a lot of money will probably get a more accurate speaker. The average accuracy for the expensive loudspeakers ($300–$500 each) is 89; for medium-priced speakers ($100–$150), 83; and for inexpensive speakers ($60–$100), 81.

Product Evaluation

Students usually come up with a fundamental question about the procedures Consumers Union uses in evaluating products: "How are ratings made?" The example of the loudspeakers presents a highly objective measure:

> Accuracy is a measure of the variation of loudness level across the audio spectrum on a scale where a "perfect" loudspeaker with a flat response has a score of 100. Differences of less than eight percentage points are not likely to be significant for most people. *(Consumer Reports 1979 Buying Guide Issue,* p. 175)

Many products do not lend themselves to such a clean technical definition of quality. A tasty example occurred in the September 1979 issue: "Which Fast Foods Are Best?" (pp. 508–13). Since teenagers are the world's greatest experts on the culinary delights described, they are eager to provide ratings on fast foods. To involve students in the problem of reducing subjectivity in ratings, it is best to have them conduct their own product evaluations before examining CU's procedures. Have students complete a rating sheet on the hamburgers, sandwiches, fish, chicken, and other items evaluated in the report (see table 23.2). If the rating sheet offers the choices Excellent, Very Good, Good, Fair, and Poor, the students' data can later be compared with CU's summary. The class will quickly be absorbed in the task of assigning and debating ratings. The discussion of differences will give the class a greater appreciation of the procedures CU used in their assessment of food quality. As described in the report, the Consumer Union's "sensory consultants" generally evaluate food in the laboratory. These consultants, whose backgrounds include home economics, nutrition, or professional food-preparation experience, usually taste products in a blind situation—they do not know what brand they are evaluating. For the report on fast foods, however,

> they tasted in the field, sampling meals at fast-food outlets in the New York New Jersey area. They submitted their usual detailed sensory evaluation of each food item. (p. 508)

Unlike the students, who are very confident about their own judgments, the CU taste experts were required to judge the foods against standards.

> There are no standards of excellence for fast foods. Therefore, we set our own criteria for the four major characteristics we always evaluate: appearance, flavor, aroma, and texture. (p. 512)

The report continues with descriptions of each of these characteristics. They are an excellent example of how CU anchors subjective judgments and provides a more objective rating of a product than the students would. Some examples include the following:

- Colors should be appropriate—french fries should be golden; chocolate shakes, chocolate brown.
- Chicken should taste and smell chickeny.
- Fish should taste and smell fresh.
- Lettuce, pickles, and the like should be crisp.

Knowing these restrictions, students are more likely to understand the results: "None of the fast-food items were rated Excellent, or even Very Good, by our sensory consultants" (p. 512).

The ratings, taken in conjunction with the nutritional information, provide for an investigation of possible influences on taste. For example, are taste ratings affected by the calories, fat, sugars, or salt in the product? Table 23.2 gives some idea of the data available in convenient form in the report. The ratings ("G" for Good and "F" for Fair) are CU's ratings. A class project based on these data should use the student ratings for more pupil involvement. My students, for example, were more positive toward

TABLE 23.2
NUTRITIONAL INFORMATION AND TASTE RATINGS FOR
EIGHTEEN FAST-FOOD MEALS

	Serving size	Calories	Fat	Carbo-hydrates	Total sugars	Sodium	Rating
HAMBURGERS							
Burger King Whopper	9 oz.	660	41 g	49 g	9 g	1083 mg	G
Jack-in-the-Box Jumbo Jack	8-1/4	538	28	44	7	1007	G
McDonald's Big Mac	7-1/2	591	33	46	6	963	G
Wendy's Old Fashioned	6-1/2	413	22	29	5	708	G
SANDWICHES							
Roy Rogers Roast Beef Sandwich	5-1/2	356	12	34	0	610	G
Burger King Chopped-Beef Steak Sandwich	6-3/4	445	13	50	0.7	966	G
Hardee's Roast Beef Sandwich	4-1/2	351	17	32	3	765	F
Arby's Roast Beef Sandwich	5-1/4	370	15	36	1	869	F
FISH							
Long John Silver's	7-1/2	483	27	27	0.1	1333	G
Arthur Treacher's Original	5-1/4	439	27	27	0.3	421	G
McDonald's Filet-O-Fish	4-1/2	383	18	38	3	613	G
Burger King Whaler	7	584	34	50	5	968	F
CHICKEN							
Kentucky Fried Chicken Snack Box	6-3/4	405	21	16	0	728	G
Arthur Treacher's Original Chicken	5-1/2	409	23	25	0	580	F
SPECIALTY ENTREES							
Wendy's Chili	10	266	9	29	9	1190	G
Pizza Hut Pizza Supreme	7-3/4	506	15	64	6	1281	G
Jack-in-the-Box Taco	5-1/2	429	26	34	3	926	G

fast foods, awarding one Excellent, eight Very Goods, six Goods, and one Fair to the foods listed.

The wider variability in student ratings may make further study of the quality issue more direct, but the Good and Fair categories listed in the report are sufficient to do some simple tests of whether quality is affected by calorie count or salt content. Table 23.3 gives the average number of calories and average milligrams of sodium for the thirteen Good foods and four Fair foods listed in table 23.2.

TABLE 23.3
AVERAGE CALORIE COUNT AND MILLIGRAMS OF SODIUM IN
FAST FOODS RATED "GOOD" OR "FAIR" IN TASTE

Rating	Average calories	Average sodium
Good	455	910
Fair	428	796

The difference in ratings seemed to be more attributable to the sodium and therefore, by implication, to the salt content of the food. Students involved in more advanced statistics courses might want to use a formal test (such as a *t* test) for the significance of the difference. Students of all levels should be prodded to produce conjectures on why the results came out the way they did. For example, should the weight of a serving be taken into consideration? Would the results be the same if milligrams of sodium *per ounce* were computed?

The two examples in this section represent extremes of measurement problems involved in the assignment of a quality indicator to products. Loudspeaker accuracy was measured with scientific instruments and represented an objective rating. The fast-food ratings were based on fallible human judgment. CU's procedures reduced the subjectivity of these judgments to produce more defensible and meaningful ratings.

Summary

Consumer Reports is a magnificent resource for statistical investigations in the high school classroom. Product evaluations of cameras, TVs, stereo equipment, food, sports equipment, and automobiles relate to the interests of the teenaged consumer. The articles illustrate good practices for defining quality. Enough data are given in the reports to give students of different academic backgrounds opportunities for the testing of hypotheses. Graphing and computing averages provide excellent activities for the less able or younger student. Those students in statistics courses can use the data for investigations requiring the computation of correlations, *t* tests, or chi squares. The teacher who uses *Consumer Reports* throughout a course will soon find that students, no matter what level, will start generating their own hypotheses, which they can then examine through the magazine.

"Does the weight of a car relate to its gas mileage?"

"Is a more powerful stereo better in quality than a less powerful one?"

"Are wood tennis rackets better than metal rackets?"

The investigation of questions like these and a study of the Consumer Union testing procedures will help the student become an intelligent user of statistics.

24. Applications of Statistics and Probability to Genetics

Hunter Ballew

How does it happen that children inherit certain characteristics from their parents? What determines exactly what characteristics each child will inherit? Why is it, in some cases, that two brown-eyed parents have all brown-eyed children, and in other cases two brown-eyed parents have some brown-eyed children and some blue-eyed children?

These can be very complicated questions, and they require the entire science of genetics for a complete answer. However, some of the simpler principles of genetics can be grasped through the application of statistical techniques and mathematical models to life situations. For example, Gregor Mendel first discovered and understood the laws of heredity by gathering and organizing with painstaking care a mass of data on pea plants. Mendel followed up the statement of his laws by comparing his data to a probability model. After studying the relationships between his data and the model, he was able to develop a plausible explanation of how characteristics are passed from parents to offspring. Although Mendel's experiments took place in the middle of the nineteenth century, the results of those experiments form the basis of the modern-day science of genetics.

Most curriculum experts and learning theorists seem to agree that the application of mathematics to real-life situations helps students master skills and understand concepts. Mendel's experiments concerning inherited characteristics in plants can be drawn on to provide applications in junior high school mathematics and science classes in two categories:

1. Basic statistical techniques of gathering and organizing data
2. The use of probability models to explain observed patterns

Statistical Aspects of Mendel's Experiments

Gregor Mendel, an Austrian monk and teacher, was interested in characteristics inherited by offspring from their parents. In particular, he was intrigued by his observation that when two unlike plants were crossbred, a characteristic of one of the parents would sometimes disappear entirely in the first generation of offspring, only to reappear in some of the plants of the second generation. For example, he had crossed a tall pea plant with a dwarf pea plant, and the seeds from this crossing had produced plants that were all tall. The characteristic of dwarfness had disappeared completely in the first generation of offspring. However, when Mendel planted the seeds harvested from the first-generation offspring, he found that some of the second-generation offspring were tall and some were dwarf.

Mendel began to wonder whether one of the opposing characteristics from two different types of parents would always disappear in the first generation of offspring after a hybrid crossing and whether the missing characteristic would always reappear in the second generation. If so, would the missing characteristic always reappear in approximately the same proportion in the second generation? He designed an experiment that would allow him to gather data that might resolve these mysteries. He was well equipped to study these problems of heredity because he was trained in both biology and mathematics. Mendel's work was the first in history to link heredity and mathematics [5].

Mendel used garden peas in his initial experiments because there were several varieties available with easily distinguishable characteristics. He identified seven pairs of traits that could be clearly described, observed, and tabulated. For example, he chose one type of pea plant that was short and bushy and paired that type with another that was tall and climbing. One kind had yellow seeds; another had green seeds. One variety had smooth seeds and another had wrinkled seeds.

Mendel first obtained seeds from plants exhibiting the seven pairs of characteristics he was interested in studying. Before he began the experiments that would produce the data he needed, he had to make sure his seeds would breed true. That is, seeds from tall plants must produce only tall plants, seeds from dwarf plants must produce only dwarf plants, and so on for all of the seven pairs of traits he planned to study. To make sure of this, Mendel planted his seeds one year, took seeds from those plants that bred true, and planted these seeds the second year. When his plants bred true the second year, he was confident that he now had pure seeds, and so he began his experiments.

In the first year of his experiment, Mendel planted his fourteen varieties in different plots. The separate plots allowed him to study each pair of characteristics one pair at a time. All other characteristics of the plants in a particular plot could be disregarded, making possible a clear and simple tabulation of results [127, p. 122].

The plants in the first year of the experiment were called the parent generation. Before the plants in the parent generation could self-pollinate, they were cross-pollinated artificially to produce hybrid pea seeds. For example, the pollen of a flower from a dwarf plant was applied to the stigma of a flower from a tall plant [127, p. 127]. Hybrid seeds from these crossings were collected and carefully labeled in relation to the type of crossing from which they resulted.

Now Mendel's curiosity was about to be satisfied. He sowed the hybrid seeds from the parent generation in the second year of his experiment so that he could see what the first generation of offspring would look like. He would soon be able to state, under controlled conditions, what plants resulting from hybrid seeds could be expected to look like. For example, if a tall plant is crossed with a dwarf plant, would the first generation offspring be expected to be tall like one parent, short like the other parent, or perhaps of medium height somewhere between the two parents? Because of his previous observations, Mendel conjectured that all plants resulting from the crossing of pure tall with pure dwarf would be tall. His conjecture was proved correct by his experimental results. In every one of the seven pairs of traits he was studying, one trait disappeared entirely in the first generation of plants after the crossing. It appeared that one member of each pair of contrasting traits had overpowered the other. This led Mendel to conclude that for each of the seven pairs of traits, one characteristic could be said to be *dominant* over the other [127, p. 133].

Mendel's plan next was to determine what would happen if the first-generation plants were allowed to self-pollinate. He wondered if the resulting seeds would produce second-generation plants with exactly the same characteristics as the first-generation plants. For example, would seeds from the first-generation tall plants always produce second-generation plants that were also tall? It might seem logical to assume that seeds from a tall plant would always produce tall plants, but Mendel's experimental results proved otherwise. He kept careful records on which seeds came from which first-generation plants, and then sowed these seeds in the third year of his experiment. This is the step by which Mendel made history. He found that some of the second-generation plants were tall and some were short. None was in between. Furthermore, he found that the ratio of tall plants to short plants was approximately three to one in the second generation. He found similar results for all seven of the pairs of traits he was studying. Table 24.1 gives, as an illustration, his results for three of the seven pairs of traits.

Table 24.1 shows how Mendel used large numbers of plants and or-

TABLE 24.1
MENDEL'S RESULTS FOR THREE PAIRS OF TRAITS

First Cross	First-Generation Plants	Second-Generation Plants
Tall × Short	All Tall	787 Tall 277 Short
Yellow Seeds × Green Seeds	All Yellow	6022 Yellow 2001 Green
Smooth Seeds × Wrinkled Seeds	All Smooth	5474 Smooth 1850 Wrinkled

ganized his information carefully in order to get an accurate description of the results. His mathematical training enabled him to analyze the results and to develop a theory to explain what happened. Mendel did not know of the existence of genes, of course, but he reasoned that there must be some factors present in the mixed plants of the first generation that transmitted the missing trait to some of the plants of the second generation. He conjectured that one of these factors (*genes,* as they are called today) must come from the male parent and the other from the female parent and that these factors must combine in some way at the moment of fertilization to determine the characteristics of the offspring.

This reasoning by Mendel illustrates the potential value of the basic statistical techniques of carefully gathering and organizing data. Using these methods, he was able to perceive patterns accurately and to state these patterns as laws of heredity with a fair amount of confidence. He next turned to some mathematical modeling to develop a possible explanation of the observed patterns.

A Model for Inherited Characteristics in Plants

After tabulating the results of his experiments, Mendel noted that the dominant characteristic appeared in the second-generation plants about three times as often as the other characteristic. This ratio was approximately the same for each of the seven pairs of characteristics he studied. From his conjecture that each trait is determined by a pair of factors, Mendel concluded that the two factors within a particular plant must have been obtained from the parents by a chance pairing at the moment of fertilization. By comparing his results to a probability model, Mendel was able to show how his observed three-to-one ratio could be expected to occur. It is remarkable that Mendel was able to use a probability model to develop a possible explanation of inherited characteristics even though very little was known in his day about what occurs physically within the sex cells when an egg is fertilized by a sperm cell.

Mendel's thinking can be studied by using modern symbolism for genes. A pair of genes that determine height can be represented by TT in a pure tall plant and ss in a pure short plant, in which each letter stands for a single gene. Suppose a pure tall plant (TT) is crossed with a pure short plant (ss). Since each parent contributes only one gene for height to the offspring and since both genes in the pure tall parent are alike, the gene from the pure tall parent must be a T. This situation is somewhat like flipping a two-headed coin. Whichever way the coin falls, the result must be a head. Similarly, the pure short parent can contribute only an s gene to the offspring.

The result of crossing a pure tall plant with a pure short plant is similar to tossing a two-headed coin and a two-tailed coin at the same time. The possible results are summarized in table 24.2. The model shows that no matter how the coins land, there is always a head on one and a tail on the other.

The possible results of crossing a pure tall plant with a pure short plant are represented in table 24.3. The structure of table 24.3 is exactly the same as that of table 24.2. The model shows that each pair of genes in the first-generation offspring is made up of a T gene from a tall parent and an s gene from the short parent. Plants with this type of genetic makeup are called *hybrid*. Mendel's observation was that all his hybrid plants of the first generation were tall. He concluded that this observation could be explained by assuming that tallness is dominant over shortness. For clarity in interpreting the symbols, capital letters are used for the dominant gene and lowercase letters for the others. The model in table 24.3 shows that all the hybrid plants of the first generation, resulting from crossing two pure plants, would be tall. This corresponds with Mendel's actual results.

TABLE 24.2
POSSIBLE RESULTS OF TOSSING A
TWO-HEADED COIN AND A TWO-TAILED COIN
AT THE SAME TIME

Second Coin

		t	t
	h	ht	ht
First Coin	h	ht	ht

TABLE 24.3
POSSIBLE RESULTS OF CROSSING
PURE TALL PLANT WITH A
PURE SHORT PLANT

Short Plant (ss)

		s	s
	T	Ts	Ts
Tall Plant (TT)	T	Ts	Ts

Now it is appropriate to examine the possible results of crossing two hybrid tall plants. Table 24.3 shows the genetic structure of each offspring of a pure tall plant and a pure short plant to be Ts. Table 24.4 shows the possible results of crossing a Ts with another Ts. The model provided by table 24.4 shows that three of every four plants in the second generation can be expected to be tall. Also, the shortness characteristic, which disappeared entirely in the first generation, can be expected to reappear in the second generation at the rate of one in every four plants. These expected results correspond closely to the results actually found by Mendel in his experiments.

TABLE 24.4
POSSIBLE RESULTS OF CROSSING TWO HYBRIDS

Female Plant (Ts)

	T	s
T	TT	Ts
s	Ts	ss

Male
Plant (Ts)

The male can contribute either a T gene or an s gene to the offspring, and the female can also contribute either a T gene or an s gene. Chance determines which combinations are made. Table 24.4 shows the four possible combinations, and all combinations are equally likely. However, three of the possibilities contain T genes, and since T is dominant over s, the probability of getting a tall plant in the second generation is 3/4, and the probability of getting a short plant is 1/4.

Table 24.4 is similar in structure to a table showing possible results of tossing two coins. The only difference between the gene model and the coin model is that the idea of dominance is missing in the latter.

Models for Inherited Human Characteristics

Many human characteristics, such as eye color, hair color, sex, ear lobe type, and tongue type, are inherited [95, pp. 44–45]. Some of these characteristics can be illustrated and studied with relatively simple probability models. One such characteristic is tongue rolling. Some people can roll their tongues into a U-shape, and others cannot. Rolling is dominant over nonrolling. In table 24.5, R stands for a rolling gene and n stands for a nonrolling gene. Each offspring from this union has a probability of 1/2 of being a roller and each offspring also has a probability of 1/2 of being a nonroller.

TABLE 24.5
POSSIBLE RESULTS RELATING TO TONGUE ROLLING

Female Parent (nn)

	n	n
R	Rn	Rn
n	nn	nn

Male
Parent (Rn)

Probability in the determination of sex

Just as it cannot be told in advance whether a flipped coin will land heads or tails, neither can it be determined with certainty whether the next child a couple has will be a boy or a girl. However, it can be safely assumed that of all babies born during a specified period of time, about half will be boys and about half girls. Actually, a few more boys are born than girls, but the figures are close enough so that a one-to-one ratio can be assumed for most practical purposes.

Units called genes are responsible for transmitting characteristics from parents to offspring, and genes are located on bodies called *chromosomes*. Chromosomes are located in cells that form the structure of living things.

Every cell in the human body, except for the sex cells, contains twenty-three pairs of chromosomes. The sex cells have twenty-three *single* chromosomes instead of twenty-three *pairs* of chromosomes. Then, when fertilization occurs, a male cell joins a female cell, and the twenty-three single chromosomes in the male cell join the twenty-three single chromosomes in the female cell to form twenty-three pairs of chromosomes. As a result, the cells of the offspring then have twenty-three *pairs* of chromosomes each, just as the cells of the parent did.

One of the twenty-three pairs of chromosomes in each cell is a pair of sex chromosomes. The female has a pair of identical, rod-shaped sex chromosomes called X chromosomes. The corresponding pair of chromosomes in the cells of the male are different. One member of the pair is an X chromosome like that in the female, but its mate, instead of being rod-shaped, is bent like a hook. Its shape is similar to a Y, and so it is called a Y chromosome.

Since the sex cell of the female (the egg) contains only single chromosomes instead of pairs of chromosomes, the sex chromosome found in each egg would always be a single X chromosome. The male sex cell (or sperm), however, could contain *either* a single X chromosome *or* a single Y chromosome. About half the sperm cells contain an X chromosome, and the other half contain a Y chromosome. The sex of an individual is normally determined at the time of fertilization [82, p. 216].

If the sex chromosomes of the female are denoted by XX and the sex chromosomes of the male by XY, a model showing the possibilities for the sex of offspring can be shown as in table 24.6. The structure of the model is similar to that shown in table 24.5. Sex, like the tongue-rolling trait, is determined by chance. If a sperm with an X chromosome happens to be the one that fertilizes the egg, the sex of the offspring will be female.

TABLE 24.6
POSSIBILITIES OF SEX OF OFFSPRING

Female (XX)

	X	X
X	XX	XX
Y	XY	XY

Male (XY)

If a sperm with a Y chromosome happens to be the one that fertilizes the egg, the sex of the offspring will be male. The model in table 24.6 shows that half the children born can be expected to be boys and half can be expected to be girls.

Sex-linked inheritance characteristics

Some people cannot distinguish certain colors or combinations of colors. Some types of color blindness are red, green, red-green, blue-yellow, and complete color blindness. Only about 1 percent of women are color-blind, but somewhere between 5 and 8 percent of men are color-blind [55, p. 486]. It has been observed that a color-blind father may have a daughter with normal vision, and then this daughter may have a son in whom the trait of color blindness reappears after having been missing for a generation. Inherited characteristics that behave in this way are called *sex-linked* characteristics.

The occurrence of sex-linked characteristics can be explained by assuming that X chromosomes contain genes that transmit certain characteristics and that Y chromosomes do not contain genes that transmit these characteristics. These genes will express themselves in male offspring even though they are not dominant because Y chromosomes, which males get and females do not get, contain no genes to prevent the expression of these sex-linked traits [81, p. 187]. A female would be color-blind only if the gene for color blindness is carried by both of her X chromosomes.

Models can be constructed to depict probabilities of color blindness occurring in offspring. In these models, X denotes a chromosome with a normal gene for color vision, and X′ denotes a chromosome with a gene for color blindness. Normal color vision is dominant over color blindness. The following combinations are possible in regard to color blindness:

XX: Normal female
XX′: Carrier female
X′X′: Color-blind female
XY: Normal male
X′Y: Color-blind male

Since there are three combinations relating to color blindness that could exist in females and two combinations that could exist in males, there are $3 \times 2 = 6$ possible mating combinations. Tables 24.7 and 24.8 show two of these six possibilities. It is possible to conclude from table 24.7 that none of the offspring from this union would be color-blind. All the female offspring, however, would be carriers of the gene for color blindness.

Table 24.8 shows that half the males born of the union of a normal male and a female carrier can be expected to be color-blind, and half the females can be expected to be carriers of the gene for color blindness. Reflection

TABLE 24.7		TABLE 24.8	

POSSIBLE COMBINATIONS OF OFFSPRING RESULTING FROM A COLOR-BLIND MALE AND A NORMAL FEMALE

POSSIBLE COMBINATIONS OF OFFSPRING RESULTING FROM A NORMAL MALE AND A CARRIER FEMALE

Normal Female (XX)

	X	X
X'	XX'	XX'
Y	XY	XY

Color-blind Male (X'Y)

Carrier Female (XX')

	X	X'
X	XX	XX'
Y	XY	X'Y

Normal Male (XY)

and the construction of other tables would show that female offspring could be color-blind only if the father is color-blind *and* the mother is either color-blind or a carrier of the gene for color blindness.

Conclusion

Illustrations have been given to show how basic statistical techniques can be used to gather data for forming possible explanations for observed results. Illustrations have also been given of the use of elementary probability models to help determine how likely it is that a particular event will occur. Instruction can be designed to reinforce and extend these skills in students. For example, after introductory material, students might be asked to do such projects as these:

1. Gather and organize data on the presence or absence of the tongue-rolling trait in members of the class, their siblings, and their parents.

2. Given that the gene for brown eyes is dominant over the gene for blue eyes, construct a probability model to show the expected eye color in children whose father has pure brown eyes and whose mother has blue eyes.

3. If a hybrid tall plant is crossed with a dwarf plant, what is the probability that one of the offspring will be tall?

4. Construct a probability model showing the various characteristics relating to color blindness that could result in children whose father is color-blind and whose mother is a carrier of the gene for color blindness.

5. What are the possible results if a pure tall pea plant with yellow seeds is crossed with a hybrid tall pea plant with green seeds?

Probability and statistics provide flexible instructional material because quite simple examples can be used to introduce the topics, and there are no real upper limits to the complexity of the material. Therefore, the subject can be challenging to students at many different levels.

25. Activities in Inferential Statistics

Barry V. Kissane

WHEN statistics courses available to all students are examined in relation to the arguments advanced for teaching them, some serious discrepancies are evident. The majority of school courses focus on two aspects of statistics:

1. The graphical representation of data and the interpretation of such representations
2. Simple numerical procedures (e.g., the calculation of mean, mode, median) and some attempt at interpreting their results

Thus statistics seems to have become the science of drawing pictures and doing sums! The interpretative aspects have tended to be minimized in practice. A consequence has been the inability of the average citizen to appreciate the need to interpret commonplace statistics carefully, let alone realize the true significance of the discipline. But there is a more serious problem than this.

The major disadvantage associated with the limited view of statistics normally experienced by students is that they fail to appreciate the very fundamental concerns of the subject. Whether statistics is viewed as "the science concerned with the collection, analysis, and interpretation of numerical information," "the study of data in context," "the study of variation," "the study of error," or some other impossibly terse description, there is no doubt that its major contribution to our understanding of the world rests on its *inferential* rather than its *descriptive* potential. If we divide our study of the world into two parts—one concerned with deterministic phenomena and one concerned with probabilistic phenomena—the second is at least as important as the first in everyday affairs. It is through the study of probabilistic phenomena that statistics makes its contribution to the world. It is a sobering thought that the vast majority of students do not realize this, despite their contact with courses labeled "statistics."

An examination of the daily newspapers reveals many situations that are mysterious to those with no knowledge of inferential statistics. Gallup polls are a good example. The press may report the findings of a poll—the average household expenditure on recreation, for instance. Although the findings relate to all households in the United States, a very much smaller number of households may have been used to gather the data. In statistical terms, the mean household expenditure is a population parameter to be estimated. The information from the sample of households provides a statistic to estimate the parameter. Provided the sample is randomly chosen (a difficult task that must be very carefully handled), a statistician can control the size of the errors of estimation involved by adjusting the sample size. The size of the likely errors is determined by the sample size, regardless of the population size. The larger the sample size is, the smaller the likely error will be. The error associated with sampling only a few thousand households to make inferences about many millions of households will thus be quite small. Estimates that prove to be considerably inaccurate usually occur because of a failure to select the sample appropriately and not generally because the sample is too small.

Newspapers also contain other applications of inferential statistics. Tomorrow's weather is predicted based on certain statistics collected. A medical report suggests that a new drug for aiding those with a particular disease has been discovered; statistics has been used to predict its effectiveness. An ecologist, aided by statistics, warns of the imminent threat to a particular species. A company advertises a twelve-month guarantee on its television sets; its quality control department will have used inferential statistics to decide on this figure.

Possible Activities

Courses that involve inferential statistics can be developed for high school students. Insofar as possible, these courses should have a strong practical focus, allowing students to interact with the data in a purposeful way. The activities undertaken should give students a good awareness of the fundamental ideas in statistics and the ways in which statistics is used. Two major aspects of such a program might be these:

1. Discussions and illustrations of how and why statistics is used in real problems of interpreting data. Some of the best sources of ideas here are the publications of the Joint Committee of the American Statistical Association and National Council of Teachers of Mathematics: *Statistics: A Guide to the Unknown* [131] is an excellent book of short readings about applications of statistics; *Statistics by Example* [85] is a series of four books (together with teacher's guides), each consisting of activities involving data analysis. These publications are all pitched at the right level and should form part of the foundation of course design.

2. Clear conceptual developments of major inferential ideas such as *distribution, population, sample, parameter, statistic, sampling distribution,*

standard error, and *confidence interval.* It is crucial that these concepts be introduced in an understandable way so that future courses can build on the students' early experiences with these ideas. It is also important that the activities used be interesting to both the students and teacher.

We shall now briefly examine two activities that I have successfully used with students in a first course in statistics, where the emphasis was on the ideas rather than on the rigor of mathematical statistics. Although the students were college students, many had no more than a high school background in mathematics and had not encountered mathematics for several years. Consequently, I am confident that these activities (perhaps with some modifications) can be successfully used with high school students. Prior to these activities, students had some experience with descriptive uses of means, variances, standard deviations, and the normal distribution. They also had some slight understanding of the concept of probability as long-run relative frequency.

Sampling a population

The first activity is based on an activity sheet on which students are instructed to select two samples from a given finite population. (See Activity 1.) Two central inferential ideas can be approached with this activity:

1. Given a particular population, what can be said about a sample chosen at random from the population?

And more important in practical terms,

2. Given a particular sample, what can be said about the population from which the sample has been (randomly) chosen?

Through this activity, students should get a strong intuitive feel for the important results:

- The means of a set of samples cluster around the mean of the population.
- The larger the sample, the closer the clustering.
- If a single sample mean is used as an estimate of the population mean, larger sample sizes lead to smaller likely errors.

I shall never forget the incredulity my first statistics teacher generated when he revealed the extraordinary fact that not only could statisticians make reasonable guesses about a population on the basis of a sample chosen from it but also that they could make reasonable guesses on how likely they were to be accurate to a specified degree! This very fact lies at the heart of inferential statistics and should be appreciated by all, not merely by statisticians.

The best way to use this activity is to pool the results from the whole class, at the same time ensuring that each student retains a copy of his or her samples. (Organizing the students into pairs so that they can check

Activity 1: Sampling a Population

The chart below contains 100 integer measurements arranged so that each is referenced by a unique two-digit number from 00 to 99. Please follow these steps:

1. Starting at a randomly chosen place in a random number table, read off five successive *pairs* of digits. For example,

    starting here gives 19, 21, 64, 02, 00

    3 2 19 2 1 64 02 00 6 9 7

2. Write the five corresponding numbers derived from the chart below in the five spaces labeled "Sample." (In the example above, these would be 27, 24, 23, 19, and 21.)

3. Calculate the mean of your sample and record it in the space labeled "Sample mean."

4. Repeat steps 1, 2, and 3 using ten successive pairs of digits in step 1 to get a sample of size 10.

5. After checking your calculations, make a copy for yourself of the slip below. (Bring the copy to the next class.)

6. Hand in the slip below to your teacher before you leave.

7. Before the next class, use a calculator to find the mean and variance of the whole set of 100 integer measurements.

	0	1	2	3	4	5	6	7	8	9
0	21	25	19	22	26	14	18	24	21	23
1	22	22	12	17	17	13	15	30	21	27
2	22	24	20	28	23	26	30	25	22	23
3	26	25	22	24	21	29	11	26	20	20
4	20	28	21	16	22	22	19	31	18	24
5	18	23	19	16	15	29	17	16	19	23
6	37	20	27	28	23	19	25	28	24	14
7	20	24	30	26	29	27	24	21	18	24
8	25	23	27	35	33	31	16	25	27	18
9	26	19	17	28	21	13	23	25	20	18

NAME _____

Sample of size 5

Sample mean

Sample of size 10

Sample mean

each other's calculations is worthwhile.) Each student then has access to three pieces of information:

1. His or her two samples
2. The distribution of the whole-class samples, pooled together by the teacher
3. The population itself

Many important ideas can be illustrated and extracted from this information at a level appropriate to the students. These include the following:

1. The *population* of interest and the *parameters* characterizing it. It is unusual to have access to the entire population; in fact, there is little to be gained of a practical nature from sampling a known population. The population has been quantified—an important point to make. The teacher may prefer to describe the population in substantive terms (e.g., the breaking strain of roof tiles, the number of years of residence in a particular suburb, the number of hours spent on homework in a week) in the interests of realism or motivation. The teacher may also prefer to discourage students from calculating population parameters until late in the discussion. The best way to do this is to delete step 7 from the activity sheet.

2. The idea of a *sample* as a subset of a population used to provide information about the population. Students have chosen samples at random. It is necessary to do this both to eliminate subconscious bias and to validate inferential procedures. The *statistics* arising from the sample are sources of information about the parameters characterizing the population. Each student should imagine that he or she has only one sample ($n = 5$) available. Why use a sample at all? Perhaps the population is inaccessible, expensive to access, or, as in the example of the roof tiles, destroyed by measurement.

3. The idea of an *estimate* as an informed guess. Ask students to estimate the population mean on the basis of their sample. They will use their sample means. Different students will have different estimates.

4. The concept of a *sampling distribution* of a statistic. The set of class means forms only a part of the sampling distribution in this activity. (The next activity illustrates some complete sampling distributions.) However, if the teacher displays the class set of means for the samples of size 5 on the chalkboard, the point is graphically made—although each sample may be expected to be quite different from other samples, some semblance of uniformity is present. The distribution of sample means will be clustered around 22 with a smaller spread than the original population. The distribution of sample means for samples of size 10 will be even more "bunched up."

5. The concept of *expected value* flows from these types of observations: what can we expect the mean, say, of a sample to be? The class or the

teacher might then calculate the mean and variance of the class set of means of samples of size 5. This is an appropriate point to reveal the population mean (22.5) and variance (25.33) for comparison with the class sampling distribution. The mean of the class means will be close to μ = 22.5 and the variance of the class means will be close to $\sigma^2/5$ = 5. Students should realize that an individual sample mean is a reasonable estimator of the population mean. If the students examine the class distribution of means of their samples of size 10, they will see that the distribution is clustered around the population mean but with a smaller variance (about $\sigma^2/10$ = 2.5). Intuitively, a larger sample will appear to provide a better estimate than a smaller sample because the results are more clustered.

Many other important ideas (some quite sophisticated) can be illustrated with this class activity, depending on the curriculum to be followed. (These may not all be illustrated in one class session, of course. The class may return to this activity at appropriate points in the program.) Briefly, some of these ideas include the following:

- The *standard error,* or the standard deviation of the sampling distribution. For the means of samples of size n, this is $\sigma/\sqrt{n}$, which decreases as n increases.

- A *confidence interval* can be constructed about the estimate of the population mean by each student using the sample mean and an estimate of the standard error. The hazy notion of confidence can be dramatically illustrated by examining the proportions of students whose confidence intervals actually contain the population mean.

- The *normal distribution* arises in the context of the central limit theorem as the limiting distribution of the sample means.

- *Student's t distribution* can be introduced to compensate for the fact that the population variance is unknown, necessitating the use of an estimate.

- *Hypothesis testing* can be illustrated by actually testing a hypothesis and comparing students' decisions on the hypothesis. An appropriate hypothesis might be μ = 20. Inferential errors might also be introduced in the same context.

This type of approach lends itself to many activities in inferential statistics. The basic idea is that a population exists and the students select samples from it. The samples are then used to provide a basis for inferring something about the population. If the population is small enough, they can compare their estimates directly with population values and also consider the properties of the whole-class distribution of estimates. If the population is large, the teacher can reveal population parameters at the appropriate (dramatic) moment. The whole process lends itself very well to using a computer either to store the population or to select the samples, or both. For those further interested in this idea, the excellent course devised by Hodges, Krech, and Crutchfield [57] will make fascinating reading. For a more complete description of the activity outlined here, see [72].

The Central Limit Theorem

Activity 2 is designed to allow students to explore in some detail the consequences of selecting random samples from a population—an activity that lies at the heart of statistical reasoning. The important outcome for students is that this leads to predictable results. In particular, the results from the previous activity concerning means of samples are again major outcomes. In addition, the normal shape of the sampling distribution arises quite naturally. Although the activity involves some mathematical derivation, it is easily modified to suit a variety of situations. For example, some may prefer to focus on graphs of the relative frequency distributions and not devote much time to the concept of mathematical expectation. The way in which the sampling distribution approaches normality is truly remarkable, even to many who are already aware of the result. In fact, I know of no better way to stimulate a study of the normal distribution.

The results in the worksheets follow from the remarkable central limit theorem, which might be stated as follows:

1. The mean of the distribution of means of random samples of size n from a population is equal to the population mean, μ.

2. The variance of the distribution of means of random samples of size n from a population equals σ^2/n, where σ^2 is the population variance.

3. If n is sufficiently large, then the distribution of means of random samples of size n from a given population is approximately normal with parameters μ and σ^2/n.

Depending on the level of the students, teachers may wish to present a formal statement of this theorem. Some teachers, of course, will prefer to leave it unstated, giving stronger emphasis to the intuitive feel that arises from the worksheets.

Other activities

If students have the necessary background, they might use populations of a different kind for inferential activities of this sort. A good example here is the distribution of the sum of two dice, where the population is best stated as a probability distribution. Samples can be selected by actually tossing dice rather than using random number tables. The same points that were raised earlier about means of samples of various sizes can be raised again.

Activity 2: The Central Limit Theorem

A central question in many statistical situations is, "How are the statistics arising from samples chosen at random from a population related to the population parameters?"

A good way to answer this question is actually to examine samples from a small population. Suppose we have a population consisting of four scores only:

$$\{2, 2, 3, 5\}$$

Here, although two of the scores are identical, we shall think of them as being distinct.

Useful parameters for describing a population are the *mean* and the *variance*, defined and computed as follows:

The population mean:
$$\mu = \frac{\sum_{i=1}^{4} x_i}{4} = \frac{2 + 2 + 3 + 5}{4} = 3$$

The population variance:
$$\sigma^2 = \frac{\sum_{i=1}^{4} (x_i - \mu)^2}{4}$$

$$= \frac{(2-3)^2 + (2-3)^2 + (3-3)^2 + (5-3)^2}{4}$$

$$= 1.5$$

Now, if we select random samples (with replacement) of size $n = 2$ from this population and find the mean of each sample, the sixteen possibilities are these:

Means, $\bar{x}$, of samples of size 2

First choice \ Second choice	2	2	3	5
2	2	2	2.5	3.5
2	2	2	2.5	3.5
3	2.5	2.5	3	4
5	3.5	3.5	4	5

If the samples are chosen at random, each of the sixteen sample means above will occur, in the long run, the same proportion of the time. That is, each sample is equally likely. So the probability of occurrence, π_i, of each sample mean is the same (1/16). We thus have a theoretical distribution of the means of samples of size 2 selected at random from our population.

Since some of the sample means are the same size, this distribution can be collapsed into the following.

Sample Mean	$\bar{x}_i$	2	2.5	3	3.5	4	5
Probability	π_i	$\dfrac{4}{16}$	$\dfrac{4}{16}$	$\dfrac{1}{16}$	$\dfrac{4}{16}$	$\dfrac{2}{16}$	$\dfrac{1}{16}$

To help interpret these results, notice that if we were to draw a very large number of samples of size 2 from the population above, we would expect 2/16 of the samples to have a mean of 4, 4/16 to have a mean of 2.5, and so on.

The mean of this distribution is

$$\mu_{\bar{x}} = \sum_{i=1}^{6} \pi_i \bar{x}_i = \frac{4}{16}(2) + \frac{4}{16}(2.5) + \ldots + \frac{1}{16}(5) = 3.$$

(Notice that this is the same as the population mean, μ).

The variance of our sampling distribution is

$$\sigma^2_{\bar{x}} = \sum_{i=1}^{6} \pi_i (\bar{x}_i - \mu_{\bar{x}})^2$$

$$= \frac{4}{16}(2-3)^2 + \frac{4}{16}(2.5-3)^2 + \ldots + \frac{1}{16}(5-3)^2 = 0.75.$$

(Notice that this is exactly half the population variance, σ^2.)

Now, complete the worksheets that follow.

Conclusion

If we are to educate citizens to understand their world more fully, some time needs to be spent acquainting them both with examples of situations in which inferential statistics can be gainfully employed and with some intuitive concept of what sorts of ideas lie at the basis of these applications. These concepts, together with some experiences at interpreting descriptive statistics and a healthy skepticism about any data collected by anybody, will appropriately arm students to meet the myriad situations they will confront in which statistics plays a vital role.

Worksheet 1

Table 25.1 lists each of the sixty-four possible samples of size 3 that can be selected from the population $\{2, 2, 3, 5\}$.

TABLE 25.1
SAMPLES OF SIZE 3

Sample			Sample Total	Sample Mean	Sample			Sample Total	Sample Mean
2	2	2	6	2.000	3	2	2	7	2.333
2	2	2	6	2.000	3	2	2	7	2.333
2	2	3	7	2.333	3	2	3	8	2.667
2	2	5	9	3.000	3	2	5	10	3.333
2	2	2	6	2.000	3	2	2	7	2.333
2	2	2	6	2.000	3	2	2	7	2.333
2	2	3	7	2.333	3	2	3	8	2.667
2	2	5	9	3.000	3	2	5	10	3.333
2	3	2	7	2.333	3	3	2	8	2.667
2	3	2	7	2.333	3	3	2	8	2.667
2	3	3	8	2.667	3	3	3	9	3.000
2	3	5	10	3.333	3	3	5	11	3.667
2	5	2	9	3.000	3	5	2	10	3.333
2	5	2	9	3.000	3	5	2	10	3.333
2	5	3	10	3.333	3	5	3	11	3.667
2	5	5	12	4.000	3	5	5	13	4.333
2	2	2	6	2.000	5	2	2	9	3.000
2	2	2	6	2.000	5	2	2	9	3.000
2	2	3	7	2.333	5	2	3	10	3.333
2	2	5	9	3.000	5	2	5	12	4.000
2	2	2	6	2.000	5	2	2	9	3.000
2	2	2	6	2.000	5	2	2	9	3.000
2	2	3	7	2.333	5	2	3	10	3.333
2	2	5	9	3.000	5	2	5	12	4.000
2	3	2	7	2.333	5	3	2	10	3.333
2	3	2	7	2.333	5	3	2	10	3.333
2	3	3	8	2.667	5	3	3	11	3.667
2	3	5	10	3.333	5	3	5	13	4.333
2	5	2	9	3.000	5	5	2	12	4.000
2	5	2	9	3.000	5	5	2	12	4.000
2	5	3	10	3.333	5	5	3	13	4.333
2	5	5	12	4.000	5	5	5	15	5.000

Answer these questions; you will need your calculator for some of them:

1. Why are some of the samples apparently identical?
2. Collapse the table into a distribution of sample means in the same way as was done for samples of size 2. (You should find nine different sample means.)
3. Find the mean of the distribution of sample means.
4. Find the variance of the distribution of sample means.
5. Compare your answers to questions 3 and 4 with the results for samples of size 2 and with the population values.
6. Guess what the answers to questions 3 and 4 would be if $n = 5$.

Worksheet 2

Although our population contains only four elements, there is no reason why a sample of size greater than 4 cannot be chosen from it. (Of course, we would not do this in practice, but it will help our discussion to do so here.) Table 25.2 contains a collapsed distribution of the means of all possible samples of size 5 (1024 of them).

TABLE 25.2
SAMPLES OF SIZE 5

Sample Total	Sample Mean	Frequency of Occurrence	Probability
10	2.0	32	.0313
11	2.2	80	.0781
12	2.4	80	.0781
13	2.6	120	
14	2.8	170	
15	3.0	121	
16	3.2	120	
17	3.4	125	
18	3.6	60	
19	3.8	50	
20	4.0	40	
21	4.2	10	
22	4.4	10	
23	4.6	5	
25	5.0	1	.0010

Use your calculator to help you complete the table and answer these questions:

1. What is the mean of this distribution of sample means?
2. What is the variance of this distribution of sample means?
3. Compare your results with the population mean and variance and with previous results.
4. Was your guess in question 6 of Worksheet 1 correct?

Worksheet 3

As well as examining statistics such as the mean and standard deviation, it is also valuable to look at the shape of a distribution. Table 25.3 contains the distribution of means of samples of size 15 from our population, with the probabilities and cumulative probabilities correct to four decimal places.

The shape of this distribution can be examined by plotting each sample mean against its probability of occurrence. The resultant graph is shown in figure 25.1.

1. This distribution looks like a normal distribution in shape. You could use the cumulative probabilities listed to check the close-

TABLE 25.3
SAMPLES OF SIZE 15

Sample Mean	Probability	Cumulative Probability	Sample Mean	Probability	Cumulative Probability
2.0000	.0000*	.0000*	3.5333	.0206	.9578
2.0667	.0002	.0003	3.6000	.0147	.9725
2.1333	.0008	.0011	3.6667	.0101	.9826
2.2000	.0020	.0030	3.7333	.0068	.9894
2.2667	.0042	.0072	3.8000	.0043	.9937
2.3333	.0081	.0153	3.8667	.0027	.9964
2.4000	.0136	.0289	3.9333	.0016	.9980
2.4667	.0211	.0500	4.0000	.0009	.9989
2.5333	.0307	.0807	4.0667	.0005	.9995
2.6000	.0414	.1221	4.1333	.0003	.9997
2.6667	.0524	.1745	4.2000	.0001	.9999
2.7333	.0636	.2381	4.2667	.0001*	.9999*
2.8000	.0727	.3108	4.3333	.0000*	1.0000*
2.8667	.0790	.3898	4.4000	.0000*	1.0000*
2.9333	.0828	.4726	4.4667	.0000*	1.0000*
3.0000	.0827	.5553	4.5333	.0000*	1.0000*
3.0667	.0790	.6343	4.6000	.0000*	1.0000*
3.1333	.0732	.7075	4.6667	.0000*	1.0000*
3.2000	.0649	.7724	4.7333	.0000*	1.0000*
3.2667	.0552	.8276	4.8000	.0000*	1.0000*
3.3333	.0457	.8733	4.8667	.0000*	1.0000*
3.4000	.0363	.9096	5.0000	.0000*	1.0000*
3.4667	.0276	.9372			

Mean of sampling distribution of means   = 3.0000
Variance of sampling distribution of means = 0.1000
*These figures are rounded to four decimal places.

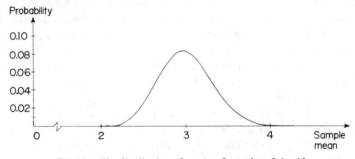

Fig. 25.1. The distribution of means of samples of size 15

ness of the sampling distribution to a normal distribution. Can you see how?

2. Use tables to check that a normal distribution has approximately 68 percent of the observations within one standard deviation of the mean.

3. For the sampling distribution (n = 15), use table 25.3 to see what proportion of the sample means can be expected to lie within one standard deviation of the mean. Compare your answer with question 2.

4. Draw graphs of the sampling distributions of means examined earlier. Compare these.

26. Statistical Inference in Junior High and Middle School

Walter J. Sanders

THE popular notion of statistics is usually limited to such "stats" as those kept on athletes (batting average, earned run average, or shooting percentage, for instance) or to such similar summaries of data as average rainfall, mean temperature for the month, and so on. To compute such statistics, all one needs to be able to do is count or add, and divide. Thus, a baseball player who is at bat fifty times and hits safely seventeen times has a batting average of 17 ÷ 50, or .340. (We usually ignore the decimal point and say the player is batting 340.) Note that there is no uncertainty here; the answer is precisely .340.

Perhaps less commonly understood is how such statistics are useful. Certainly a player who is batting 400 is in a better position to negotiate a new salary than a player with a 200 average. But more important to the team is how such averages help the manager decide which player to use as a pinch hitter or whether to call for a hit-and-run play, bunt, or whatever, depending on which players are on base, at bat, on deck, and so forth. But putting in a hitter with a 400 average will not guarantee the manager that the player will get a hit. In fact, a hitter with an average of 200 may get a hit, and a 400 hitter may strike out. The decisions a manager makes are not sure things; they are made in the face of uncertainty and are subject to the variability of chance.

The ability to compute averages from data is one of several skills helpful in organizing data that has been collected. In a sense, the average is a condensed description of the data. Other common descriptors of data are the mode, median, and standard deviation. The study of ways to collect, organize, and describe data is called *descriptive statistics*.

Statistical experiences currently available to junior high and middle school students are almost exclusively concerned with descriptive statistics. For example, students might go to a busy intersection and collect all sorts of data on the cars they see passing: color, number of passengers, home state of license, and so forth. The data are sorted and presented as a histogram or bar graph, and the mean, median, and mode for each set of data are discussed. In some cases, a standard deviation may even be computed. Seldom are students asked to draw conclusions that involve the uncertainty of chance from such data or to try to measure the uncertainty involved.

Making decisions and drawing conclusions from data, especially where such decisions and conclusions are uncertain because of the variability of chance, are the fundamentals of *inferential statistics*. The purpose of this article is to show how students at the junior high and middle school levels can become involved in appropriate inferential statistics experiences and to describe materials that can be used to carry out such a program.

Statistical Inference

The fundamental problem of statistical inference is to determine the way specific attributes are distributed among the members of a given group, called a *population*. (Here, *attribute* will refer to any characteristic shared by several members of a population.) For example, in the manufacture of shoes one must be concerned with the distribution of shoe sizes among all possible customers. The proportion of shoes produced of any given size will depend on the proportion of the population of potential customers who wear that size. Shoe manufacturers cannot measure the shoe size of each person in the country. However, they could determine the shoe sizes of a few people and try to predict from such a *sample* how many of each size to make. Thus, if 4 percent of the sample wear size 10B, they might assume that 4 percent of the population also wear size 10B, and make 4 percent of their total production of shoes size 10B. Unfortunately, the decision is more complicated than this. If a second sample were taken, it might turn out that this time 6 percent of the sample wear size 10B. This might very well happen because the sampling procedure involves a degree of *chance variability;* the distribution of an attribute in one sample will ordinarily be different from the distribution of that attribute in another sample.

One factor to consider in making predictions from a sample is the size of the sample. If the shoe manufacturer uses a sample of ten people, there might not be any who wear size 10B, whereas in a sample of 100 000 the percentage wearing size 10B will probably be highly representative of the population. How can an appropriate sample size be determined?

Another factor to consider is how the sample is to be selected. If the sample consists entirely of women, there will be no information on men's shoe sizes. Care must be taken to ensure that the sample represents the population. The sample must be selected in such a way that each member of the population has as good a chance of being included as any other member. A sample selected in this manner is called a *random* sample.

When a random sample of optimum size has been selected, there is still the problem of contending with the variability of the sample because of chance. In fact, the major problem of inferential statistics is to find ways to *measure* the uncertainty of conclusions about a population caused by the chance variability of the sample.

A typical problem in the field of education is to determine whether an attribute is distributed the same in two populations. Thus, to determine whether method A or method B of teaching long division is better, we begin with two equated groups of children and teach one group by method A and the other by method B. We must then decide whether the difference in the children's posttest scores is attributable to chance variation or to the superiority of one method over the other. To answer this question, we must have a way of measuring the variation that can be attributed to chance.

Sampling with Concrete Materials

Students can get an intuitive grasp of statistical inference by carrying out a variety of laboratory experiments with concrete materials. To become familiar with the reliability of sampling estimates of the distribution of attributes in a population, the students generate sample distributions using a sampling paddle and colored balls. To make a sampling paddle, sink holes into one side of a piece of wood so that when it is drawn through a population of balls, balls roll into the holes (see fig. 26.1). When withdrawn, the paddle contains a sample of the population. It will be useful to have sampling paddles of three sizes: five holes, ten holes, and fifty holes.

Prior to class, you can use the sampling paddles to select quickly a population of 2000 balls that is 4 percent red and 96 percent green (assuming the balls have already been sorted). One scoop with the fifty-hole paddle and three scoops with the ten-hole paddle will give 80 red balls (4 percent of 2000); thirty-eight scoops with the fifty-hole paddle and two scoops with the ten-hole paddle will give 1920 green balls. These are mixed in a cardboard box.

Without permitting the students to see inside the box, ask what they think the box contains. A shake of the box will lead some to guess that it contains balls. The sound may also produce guesses concerning the number

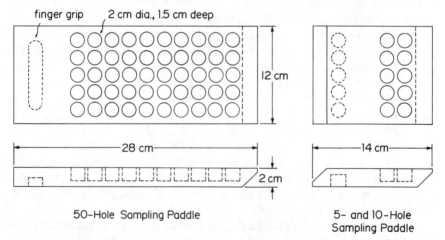

finger grip   2 cm dia., 1.5 cm deep

12 cm

28 cm   2 cm

14 cm

50-Hole Sampling Paddle   5- and 10-Hole
                                      Sampling Paddle

Fig. 26.1. Plans for 50-hole sampling paddle and for 5- and 10-hole sampling paddles for use with marbles 1.5 cm in diameter

of balls. After telling the students that there are 2000 balls in the box, you can ask them about the colors. Emphasize that at this point there can be only guesses. Then reach in and draw out one ball; let us say it is red. Ask the students what they now know for sure about the colors and also what they speculate about the colors. Of course, the only thing they can be sure of is that there is one red ball. But if the ball has been drawn randomly, they might guess that a large portion of the balls are red.

Now draw a sample of five from the box. Since there are 80 red balls, this sample could contain any number of red balls, 0 through 5. Let us say it contains 2. What predictions might the students now make? The most likely guess is that 800 (two-fifths of the 2000 balls) are red. Ask students to put limits on this estimate by asking them such questions as "Are there likely to be fewer than 750 red balls?" "More than 900?" Draw another sample of five; let us say that this sample has no red balls. Repeat the questions asked for the first sample. Be sure to raise the question of how the new estimates can be reconciled with the earlier estimates. Some students may suggest that 400, the average of 0 and 800, is now the best guess. Repeat the procedure with several more samples. Since samples have included only red and green balls, students will predict that all or nearly all must be red or green. This would be a good time to admit that there are only the two colors.

Continue by drawing samples of size ten, and then samples of size fifty. This will not only give considerably more information about the proportion of the colors but will also show that sample size makes a big difference in the accuracy of the estimates and the limitation on chance variability. It gives very little information, however, on what a *measure* of the variation might be. During the discussion, some students may accuse you of having all the red balls at one end of the box and all the green balls at the other. This is a good opportunity to let a student stir the marbles around (without looking, of course); then you can talk about randomness in the sample. You might look in the box while drawing a sample to get the students to criticize this method as one that will not produce a random sample.

Someone should now count the red and green marbles and thus establish that there are 4 percent red and 96 percent green balls in the box. Then the class can note how accurate their predictions were.

Sample Distributions

One can get an idea of how much variation there is among samples from a given population and prepare for a way to estimate the chance variation of a sample from an unknown population by drawing a number of samples from a known population and recording the results on a graph. For example, mix a population of 2000 balls that is 10 percent red and 90 percent green (200 red balls and 1800 green). Using a sampling paddle, draw a sample and count the number of red balls in the sample. Mark an X on graph paper to represent this number. Continue taking samples and recording on the graph the number of red balls in each sample until 100 samples have been taken. Typical graphs (called *sample distributions* or *empirical* sample distributions) are shown in figure 26.2 for samples of several sizes. In each instance, 100 samples were taken so that the percentage of samples with a given number of red balls could be readily determined. In figure 26.2(a), which records samples of size five, 23 of the 100 samples (23 percent) contained two red balls. In figure 26.2(b), which records samples of size ten, there were 29 samples (29 percent) with two red balls. In figure 26.2(c) and (d) none of the samples (0 percent) had two red balls.

The effect of increasing the sample size can be seen from the distributions in figure 26.2. Since the distributions with larger size samples are spread out more, it might *seem* as though the smaller the sample size, the more closely the sample distribution "narrows in" on 20 percent, the true percent of red balls in the population. That this is not true can be seen by considering the *percent* of red balls in a sample. Two red balls in a sample of five is 40 percent of the sample, whereas two red balls in a sample of fifty is 4 percent of the sample. For samples of size five, the number of red balls ranged from zero to four, which is from 0 percent to 80 percent of the sample. But for samples of size fifty, the number of red balls ranged from three to seventeen, which is from 6 percent to 34 percent of the sample. So, samples of size fifty in fact narrow in on the percentage of red balls in the population much more closely than samples of size five. Figure 26.3 shows the graphs of each of the four distributions of figure 26.2 when the

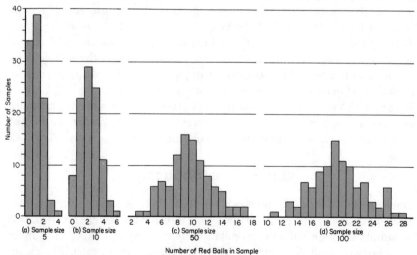

Fig. 26.2. Distributions of 100 samples of different sizes with the horizontal axis given as the number of red balls in each sample (from a population with 20 percent red balls)

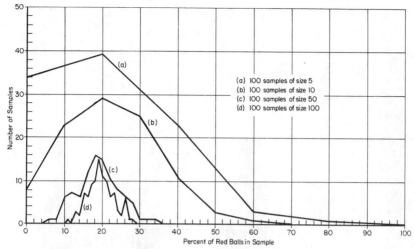

Fig. 26.3. Distributions of 100 samples of different sizes with the horizontal axis given as the percent of red balls in each sample (from a population with 20 percent red balls)

percent (rather than the number) of red balls in a sample is used. The superposition of the graphs shows that all four cluster around 20 percent, the actual percent of red balls in the population, and that the larger the sample size, the narrower the spread of the distribution.

It can also be seen that although samples of size 5 and size 10 are quite spread out, samples of size 50 are nearly as closely clustered as samples of size 100. Because of this, and since samples of size 50 are convenient to work with, they will be used in the remainder of the sampling experiments.

Sample Distributions for Different Populations

To prepare for a way to estimate the chance variation of a sample from an unknown population, make sample distributions of size fifty for populations that have 1 percent, 2 percent, and so on, up to 25 percent red balls with the rest being green. The work can be done by students. A team of three can prepare a sample distribution of 100 samples in about twelve minutes. Thus, two fifty-five minute laboratory sessions are more than adequate for six teams to prepare sample distributions of 100 samples each for the twenty-five populations. Figure 26.4 shows sample distributions for 3 percent, 6 percent, 9 percent, 12 percent, and 15 percent populations.

In preparing these distributions, students may make several observations. One is that during sampling it is common to have a *run* on one number, but if the sampling is carried out long enough, the runs are balanced out. Another observation is that the graphs do not exhibit a smooth regularity; one column may end up with more Xs than a column closer to the center.

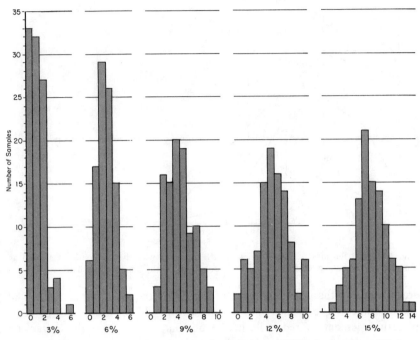

Fig. 26.4. Sample distributions for several populations of 2000 using samples of 50

A third observation is that many of the graphs appear to be skewed to the right.

Students are intuitively aware that in the 10 percent distribution samples of five red balls should be drawn more often than samples with any other number, but the distribution had more samples with four than with five. Possible reasons that students give are that there was a run on samples with four red and that later it would even out or that there had been some type of bias introduced into the sampling procedure. Finally, students are quite willing to state that on the basis of their experimentation one would expect to draw a sample with zero red balls about 1 percent of the time and samples with four red balls about 22 percent of the time. Thus, they develop an intuitive feeling for the probability of a sample having a particular number of red balls without necessarily understanding what probability is.

Estimates Using Sample Variability

We are now ready to solve a problem. Suppose a sample of 50 is drawn from a population of 2000 that has an unknown percentage of red balls. The sample contains 4 red balls. What reasonable estimates can be made about the percentage of red balls in the total population? All we know for sure is that the population contains at least 4 red balls and at least 46 nonred balls, since that is what we saw in the sample; so there are, at most, 1954 red balls. Theoretically, then, the population could contain as few as 0.2 percent red balls (4 out of 2000) or as many as 97.7 percent red balls (1954 out of 2000). Very few mixtures are ruled out!

But when we look at the sampling distributions for 1 percent through 25 percent mixtures (as represented in fig. 26.4), we see that only about one-fifth of the theoretically possible mixtures are likely to produce a sample with four red balls. The graph in figure 26.5 shows the number of samples with four red balls from each of the sample distributions. No sample with four red was drawn from any population with more than 20 percent red. This strongly suggests that the unknown population is likely to have between 1 percent and 20 percent red balls, which is a far more restrictive estimate of the unknown population than the theoretical limits of 0.2 percent to 97.7 percent. A knowledge of the variability of sampling has allowed us to narrow the limits of our estimate of the unknown mixture more than we could do otherwise.

Of course, when we say that the unknown mixture *probably* lies between 1 percent and 20 percent, we cannot be absolutely sure we are right. But it would surely be a rare occurrence for such an estimate to be wrong. If we are willing to increase the chance of being wrong, we can establish even closer limits on our estimate. For instance, we might decide to ignore all populations for which 1 percent or fewer of the samples contained four red balls. From figure 26.5 we see that that would leave us with 3 percent through 18 percent populations; so, in this instance, we estimate that the

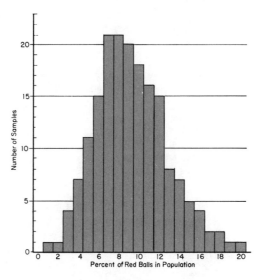

Fig. 26.5. Number of samples containing four red balls from populations of 1 percent through 22 percent red balls

unknown population contains between 3 percent and 18 percent red balls. This gives us closer limits on our estimate but also increases the likelihood that we could be wrong (we actually *had* samples with four red balls from populations we are omitting).

We can elect to ignore populations up to whatever percent of samples with four red balls that we choose. But the more populations we ignore, the greater our chances of being in error. Thus, if we decide to ignore all populations with fewer than 10 percent of the samples having four red balls, the limits of our estimate are from 5 percent to 12 percent. We have narrowed our estimate to a range of eight populations but have also increased the margin of error.

The experiments described here have been carried out with seventh- and eighth-grade classes. The procedure was tested by taking a number of samples of each of several unknown populations to see how often the population was within the limits determined above. The results of the experiment support the procedure of estimating from sample variability. In fact, when the limits were set by including all populations with one or more appropriate samples, the estimates were always correct. When the limits were set to include all populations with ten or more appropriate samples, the estimates were correct more than 70 percent of the time.

VII. Monte Carlo Techniques and Simulation

27. Monte Carlo Simulation: Probability the Easy Way

Ann E. Watkins

EACH box of a certain brand of cereal contains one of seven cards about superheroes. A student in an elementary school class wants to know how many boxes of cereal he can expect to buy before getting the entire set. Since this is a difficult problem to solve analytically, the class decides to solve it by simulation. The students put the names of the seven superheroes on slips of paper and put the slips into a box. They draw a name, write it down, and replace the slip of paper. They continue until each name is drawn at least once and record the number of boxes "purchased." Repeating the process twenty times, they find the mean number of boxes is 18.7.

▶ Time has run out in the big basketball game, and the score is a tie. However, the high school's best free-throw shooter, with a .75 average, was fouled and gets two shots after a short time-out. What, the students wonder, is the probability that she will make at least one shot and win the game? They quickly construct two spinners with paper clips (see fig. 27.1). Holding the paper clips in place with the tips of pencils and spinning the pair of spinners thirty times, they find that at least one spinner lands on "make" twenty-eight times. They relax.

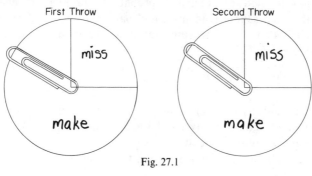

First Throw   Second Throw

miss   miss

make   make

Fig. 27.1

▶ A community college student council is considering placing the Social Club on probation. The overall campus grade point average (GPA) is approximately normally distributed with a mean of 2.4 and standard deviation of .5. The Social Club has twenty-five members and a GPA of 1.9. One member of the Social Club is a mathematics major who knows a little about normal curves. He points out to the council that 16 percent of the students at the college have a GPA as low as 1.9. The council is not impressed and decides to use the school's computer to find out how unusual it is for a randomly selected group of twenty-five students to have a GPA this low. The council has the computer randomly select twenty-five points from under a normal curve with mean 2.4 and standard deviation .5 and compute their mean. The process is repeated 100 times. If the mean GPA is 1.9 or less, the run is counted as a success; otherwise, it is a failure. Only one run is a success. The Social Club is placed on probation.

▶ The Kid has challenged Doc to a shoot-out. Both are bad shots. The probability that Doc hits the Kid on any one shot is 1/10, but the probability that the Kid hits Doc is 1/5. Doc thinks the shoot-out may be a bad idea. "Okay," says the Kid, "to even things out, you can go first and we'll alternate shots until one of us gets hit." Doc wonders if this really evens things out. He gets out his table of random numbers (a telephone directory) and asks the Kid to pick a digit between 0 and 4. The Kid picks 3. Doc picks the number 5 between 0 and 9 for himself. Starting at a random spot on the table, Doc will let each number represent a shot. If it is his turn to shoot, a 5 will represent a hit and any other digit a miss. If it is the Kid's turn, a 3 will be a hit and 0, 1, 2, or 4 will be a miss. If it is a 5, 6, 7, 8, or 9 on the Kid's turn, he will disregard it and use the next digit instead. Repeating the procedure 100 times, Doc finds he wins 37 times but gets shot 63 times. He chickens out.

These problems have been solved by a technique called *Monte Carlo simulation.*

What Is Monte Carlo Simulation?

The usual method of solving a probability problem is analysis. For example, the high school students could have reasoned that the basketball player has a 1/4 chance of missing each shot, so her chance of missing both shots would be

$$\frac{1}{4} \cdot \frac{1}{4} = \frac{1}{16} \ .$$

However, an analysis of interesting problems is often beyond the competence level of the class. Analyses of the superhero, shoot-out, and Social Club problems are quite difficult.

A second method of solving probability problems is by experiment. If they had wished to solve their problem by experiment, the high school students would have asked the basketball player to take, say, thirty pairs of free throws and then counted the number of times she missed both shots. Experimentation is frequently expensive and time-consuming. To solve their problem by experiment, the elementary school students would actually purchase boxes of cereal. Doc had additional reasons for not wishing to solve his problem by experiment.

The superhero, free-throw, Social Club, and shoot-out problems were solved by Monte Carlo simulation. Monte Carlo simulation becomes necessary when mathematical analysis is difficult or impossible and experimentation is expensive, time-consuming, or otherwise impractical. It involves finding a model for the given problem. This model is physically different and easier to operate but has the same mathematical characteristics as the original problem. The distinctive feature of the Monte Carlo method is the use of random devices such as dice, coins, spinners, or random numbers from tables, the phone book, or the random-number generator of a computer.

The theoretical basis for the Monte Carlo method is called the *law of large numbers,* which states that as a simulation is run a larger and larger number of times, the simulated estimate

$$\frac{\text{number of successes}}{\text{number of runs}}$$

becomes approximately equal to the analytical probability.

Developing your own simulations

Select a problem. Almost any probability or expected-value problem can be solved by a suitable Monte Carlo simulation. One basic type of problem involves determining the probability of a success or of a failure. The high school students were simulating a problem where a success was making one or both free throws. A success to Doc was when the Kid got hit. A success for the student council was finding a mean GPA as low as 1.9.

A second basic type of problem asks for an expected value, not for a probability. For example, the elementary school class answered the question, "How many boxes of cereal can one expect to buy before getting the entire set of superhero cards?"

The next section contains more sample problems.

Identify the essential mathematical characteristics of the problem and assign probabilities to the outcomes. First, identify the possible outcomes of each trial. For the elementary school students, one of seven outcomes was possible for each trial because for each purchase the box of cereal contained one of seven superhero cards. For the high school students, the two possible outcomes of each free throw were make and miss. In the shoot-out, the possible outcomes of each shot are Doc hit, Kid hit, and nobody hit.

The community college students have an essentially infinite number of possible outcomes, since the GPA of one student is a number between 0 and 4.

Next, determine the probability of each outcome. In the superhero example, the outcomes are assumed equally likely; so each has the probability 1/7. In the Social Club example, the probabilities are given by the position of an outcome on the normal curve. In the shoot-out, the probabilities depend on whose turn it is to shoot. If the Kid is shooting, for example, the probability of the Kid being hit is 0.

Be sure to consider whether or not the trials in the original problem are independent. The high school students assumed that the free throws were independent of each other—that success or failure on the first free throw would not change the probability of .75 of a success on the second. Because of the pressure on the basketball player, this assumption is probably not correct.

Select a random device. A random device must be selected that embodies the mathematical characteristics of the problem. The possible outcomes of the problem are matched to outcomes of the random device that have the same probability. To use his table of random numbers to generate a probability of 1/5, Doc had to ignore half the digits.

There are usually several ways of simulating the same problem. The elementary school students could have chosen a spinner divided into seven equal areas.

Determine what a run will consist of. The high school students completed a run by spinning both spinners. The run was considered a success if one or both spinners landed on "make." Doc completed a run when a hit occurred. He considered the run successful if it was the Kid who got hit. The elementary school students continued to draw slips of paper until all seven names appeared at least once. Since they were doing an expected-value problem, they recorded the number of slips drawn on each run.

Do a large number of runs. If possible, estimate how many runs of the simulation are needed to give a sufficiently accurate estimate. In general, the larger the number of runs, the more accurate the result. However, to cut the error in half, the number of runs must be quadrupled. For a problem in which the probability of a success is needed, try $1/e^2$ runs, where e is the largest error you wish to allow, and you will be 95 percent sure that your simulated estimate is within e of the analytical probability. If the high school students wanted to be 95 percent sure of knowing the probability of making at least one free throw to within .05, they would do

$$\frac{1}{(.05)^2} = 400 \text{ runs.}$$

It is difficult to determine the number of runs needed for an expected-value problem. Some suggested numbers are included with the sample problems.

The number of runs needed for good accuracy can be quite large, but if, for example, 300 runs are needed and there are thirty students, each student would do only 10 runs.

Compute the simulated answer. For probability problems, the simulated estimate is

$$\frac{\text{number of successful runs}}{\text{number of runs}}.$$

For example, Doc ran his simulation 100 times and had 37 successes; so his simulated probability is

$$\frac{37}{100} = .37.$$

For expected-value problems, the expected value is

$$\frac{\text{sum of results from all individual runs}}{\text{number of runs}}.$$

For example, the elementary school class may have drawn these numbers of boxes for their twenty runs:

20	14	27	18	17	15	19
19	19	20	16	11	15	21
15	22	20	28	12	26	

Thus, their simulated estimate is

$$\frac{374}{20} = 18.7.$$

Sample problems

This section contains additional problems that are fun to solve by Monte Carlo simulation. In parentheses following each problem are the analytical solution and then a suggested number of runs.

1. What is the probability that all five children in a family will be girls? (.03, 280)

2. If women make up 30 percent of the labor pool but a particular company has only fifteen women out of sixty-two employees, what is the probability that the company's hiring record could have occurred by chance? (.16, 325)

3. A school has overbooked its twenty-two spaces in the teachers' parking lot. It sold twenty-five permits. Assuming there is a .1 chance of any given teacher being absent on a given day, on what proportion of days will there be at least one teacher without a space? (.63, 360)

4. What is the probability that, in a group of twenty-three people, two

will have the same birthday? Ignore February 29 and assume that all 365 birthdays are equally distributed. (.5, 385)

5. How long is the average shoot-out? (6.8 shots, 145)

6. A system with six components fails if one or more components fail. The probability any given component fails is .2. What is the probability that the system fails? (.74, 300)

7. A man leaves for work at a random time between 7:00 A.M. and 8:00 A.M. His newspaper arrives at a random time between 6:30 A.M. and 7:30 A.M. What is the probability that the man will get his paper before he leaves for work? This problem can be simulated using two spinners like those in figure 27.2. (7/8, 170)

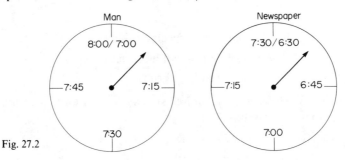

Fig. 27.2

8. How would the expected number of cereal boxes change in the super-hero example if the outcomes were not equally likely? (It increases.)

9. A student takes a ten-question, multiple-choice test with four possible answers for each question. What is her expected score if she guesses on every question? (2.5, 180)

10. If parents continue having children until they get a girl, what is their expected number of children? What proportion of their children will be boys? (2 and .5, 200)

11. A cloakroom attendant receives nine hats from nine men and gets the hats mixed up. If he returns the hats at random, what is the expected number that go to the correct owner? (1, 100)

12. A music teacher requires his students to play a piece perfectly three times in a row before they get a gold star. A student estimates that she can play "One Note Sonata" perfectly seven times out of ten. How many times does she expect to have to play "One Note Sonata" before she gets the star? (6.4, 300)

Why Teach Monte Carlo Simulation?

Although the method had been used before, Monte Carlo simulation was named and developed during World War II by a group of mathematicians including John von Neumann to solve problems that arose in the design of atomic reactors. Since then, the availability of computers has enabled industry to use the Monte Carlo method to solve a wide range of problems, including the absorption of high-energy radiation, the computation of the thermodynamic properties of various systems, the development of statistical tests, war games, and signal detection in the presence of random noise. Because it is so commonly used, students should be familiar with the technique. They should know what it can be used for and understand its limitations.

As a type of simulation or mathematical model, the Monte Carlo method teaches students how to represent a real-world system in terms of mathematical relations. As we have seen, they learn how to isolate the critical factors in a problem. Is each trial independent? Are the outcomes equally likely? Then, students must carry out operations that imitate the manner in which the real system behaves.

Probability is usually a difficult subject for students. Having been introduced to probability relatively late, they have little intuition for it or experience with it. Monte Carlo simulation is a good introduction to probability. Clarifying the assumptions students make about independent trials, the probability of outcomes, and so on, serves as a way to introduce vocabulary and definitions and is a first step to solving problems by analysis. In addition, simulation gives students a feeling of power over probability. It is a technique they can almost always use.

Teachers have found simulation useful as a way of verifying results that students are reluctant to believe from a purely analytic explanation. For example, if the probability that a baby is a boy (B) is 1/2 and the probability that it is a girl (G) is 1/2, is the probability of having two boys in a family of two children 1/3 or 1/4? That is, is the equiprobable sample space

BB   BG   GG

or is it

BB   BG   GB   GG?

Students can be convinced that it is the latter by flipping two coins with, say, heads designated a boy and tails a girl. If this is done 100 times, a distribution such as

HH	HT	TT
23	48	29

will result, indicating that these three outcomes are not equally likely.

Finally, Monte Carlo simulation is easy to do in the classroom. Most needed materials can be quickly made. If a computer is available, the Monte Carlo method is a convincing way of demonstrating its power. Above all, students find this kind of mathematics fun and interesting and enjoy trying to devise new and clever variations.

28. Using Monte Carlo Methods to Teach Probability and Statistics

Kenneth J. Travers

MONTE CARLO methods are not really new. Some authors (see, for example, [52]) date them back to biblical times. During the past forty years, however, they have been developed and used in a variety of fields of research. Hecht [54] has identified over twenty-five hundred articles published since 1960 that employed these methods, and a current issue of *Science Citation* indexes about 150 articles referencing Monte Carlo methods.

What Are Monte Carlo Methods?

The well-known mathematician S. M. Ulam formalized Monte Carlo methods and used them in his research in the mid-1940s at Los Alamos, New Mexico. The methods emerged from conversations he had there with John von Neumann, another well-known mathematician. Ulam appears to be pleased with the name "Monte Carlo," chosen, he states in his autobiography, "because of the element of chance, the production of random numbers with which to play the suitable games" [139, p. 199].

Ulam was pondering the solution, not to a problem involving dice, as were those earlier mathematicians Pascal and Fermat, but to one involving a card game. As he describes the experience in his article "Computers" [140], he was attempting to determine the fraction of all games of solitaire that could be completed satisfactorily to the last card. He was not able to determine a general solution to the problem; so it occurred to him that a useful approach to the problem would be to have a computer play out a number of games, say 100 or 200, and record the results. This, he claims, was the origin of Monte Carlo methods.

In the classroom, Monte Carlo solutions to problems typically begin with identifying appropriate models to employ (such as coins or dice) and deciding how they might be used to solve the problem. Data, such as the number of heads on the toss of three coins, are obtained, preferably with the students actually producing the data in the class. Finally, the results are tabulated and interpreted in terms of the given problem.

An example of Monte Carlo methods

Munchy Crunch Cereal offers a pen of one of six different colors in a box. Assuming equal chances of getting any of the six colors of pen with one purchase, how many boxes of Munchy

210

Crunch would one expect to have to buy to obtain the complete set of six pens?

In order to get a handle on this problem, we take our lead from Ulam. Since the solitaire problem was too difficult for him to solve by conventional, analytical methods, he conducted an experiment, as will we.

The Munchy Crunch problem is interesting to students. Who as a child has not tried to complete a set of baseball cards by buying bubble gum or a set of trinkets by purchasing popcorn? This problem is too difficult to solve by conventional methods involving probability theory for all but advanced students in a high school probability course (how many mathematics majors in college could solve it?). Using Ulam's Monte Carlo approach, however, typical eighth-grade students with no prior formal instruction in probability whatsoever can solve this problem.

We begin by obtaining guesses about the number of boxes one might expect to have to buy to get the complete set of six colored pens. Is it about fifty boxes? (Thirty-six is a favorite guess; explanations of how thirty-six is obtained are often intriguing and sometimes impressive.)

Then, we move the discussion along by a comment like, "Well, suppose there were a prize, such as a moped or a motorcycle, for the best answer. How could you get an idea about how many boxes would be required?" This typically leads someone to suggest that one could go out and start buying cereal! This is a promising suggestion, and we prepare a chart like this:

Shopping trip number	Color of pen						Number of boxes purchased
	Pink	Purple	Auburn	Rust	Turquoise	Tawny	
1	TH꜀	/	TH꜀ /	//	///	//	19

The chart is used to keep a record of how many pens of each color were acquired with the purchases. The "shopping trip" ends when a complete set of the six different pens has been obtained. Our trip ended when we got a purple pen. If we want to see whether we were lucky or unlucky on the trip, we could go on another trip and compare the numbers of boxes required as the result of two trips. Better yet, we can take many shopping trips and find the average number of boxes. Students seem to need very little persuasion to accept the notion that the more shopping trips we include in our averaging, the more confidence we can have in our answer.

The classes usually are quick to agree, as well, that we are not serious about actually buying cereal. We now search for a solution that could be found right in their own classroom. Can we find a model of the chances of buying a box of cereal and getting any one of the six pens? Usually, a member of the class suggests using a die. We can now restate our problem in terms of a die, using the chart that follows.

Shopping trip number	Color of pen						Number of boxes purchased
	1 = Pink	2 = Purple	3 = Auburn	4 = Rust	5 = Turquoise	6 = Tawny	
1	//	////	//	/	//	////	16
2	/	//	///	//	////	//	14

We now summarize the findings of the class using a table (see table 28.1), where X represents the number of rolls of a die required to obtain all faces of the die. A total of only twenty "shopping trips," or trials, was used. It is easy in a class of twenty-five students to get at least 100 trials (4 from each student).

TABLE 28.1
SUMMARY OF CLASS RESULTS

Number of cereal boxes per trip X	Frequency	Number of boxes purchased
7		
8	/	8
9		
10	/	10
11	/	11
12	/	12
13	/ / /	$13 \times 3 = 39$
14	/ / / /	$14 \times 4 = 56$
15	/ / /	$15 \times 3 = 45$
16	/ /	$16 \times 2 = 32$
17	/	$17 \times 1 = 17$
18	/	$18 \times 1 = 18$
19		
20	/ /	$20 \times 2 = 40$
	20	288

The average (mean) number of boxes of cereal purchased in twenty trips = 288/20 = 14.4.

Since this table is constructed from the results of the class activity, some discussion is worthwhile. The class is asked, for example, how small X could be. The class is soon convinced that X could be as small as six, but that would only be for very lucky people—those who get a pen of a different color with each box of cereal! Well, then, how large might X be? It could be very large—for the very unlucky person who could be imagined as having bought hundreds of boxes of cereal and still not getting that last pen! But the results of the class will soon show that, by and large, the average value is around fifteen boxes. (It can be shown analytically that the theoretical value is $6/1 + 6/2 + 6/3 + 6/4 + 6/5 + 6/6 = 14.7$.)

A simple variation of the problem poses that the Munchy Crunch Company decides to offer pens of eight different colors rather than six. How many boxes of cereal can be expected to be bought now?

At this point it is extremely useful to have access to a set of polyhedral dice. The cube has already been put to good use, and the octahedron's turn has come. Classroom sets of these dice are available from educational supply houses. However, other models for obtaining eight equally likely outcomes are available. An eight-sided roulette wheel or spinner could be carefully constructed, or one could simply use an opaque bag containing eight marbles that are different in color but otherwise identical. Put the marbles in the bag, mix them up, draw one without peeking, note its color and replace it in the bag; repeat. Another source of equally like outcomes is the computer. The following program in BASIC will do the job:

```
10  PRINT INT(8*RND(-1))+1,
20  GO TO 10
30  END
```

Here is a sample output:

3   2   3   2   6   1   2   6   8   3   4   1   4   1

Or, one could go to amost any college-level statistics book for a table of random numbers. The tables are produced in such a way that each of the ten digits, 0–9, is equally likely to appear in the list. Then, one could use the digits 1–8 and ignore 0 and 9. Presto, a table giving eight equally likely outcomes. The example below was taken from a table of random numbers.

72234   40065   24052   05658   08335

Now that we have our source of random outcomes, we proceed as before to find the number of boxes we would expect to have to buy for the complete set of eight pens. Sample results are given in table 28.2. We now call each shopping trip by its more technical name, *trial*. With a class of about thirty students, 100 trials can be obtained within a few minutes, each student conducting 3 or 4 trials.

TABLE 28.2
RESULTS FOR FOUR SHOPPING TRIPS (TRIALS)

Trial	Pens								Number of boxes of cereal
	1	2	3	4	5	6	7	8	
1	//	┼┼┼ /	/	//	┼┼┼ //	///	//	////	26
2	///	//	////	///	/	//	///	┼┼┼	23
3	//	////	/	//	///	//	////	//	19
4	┼┼┼	┼┼┼	///	//	///	/	//	┼┼┼ //	28

The mean (average) number of boxes per trial = (26 + 23 + 19 + 28)/4 = 96/4 = 24.0.

For more advanced students, it is of interest to do some generalizing. What happens to the expected number of boxes of cereal required to obtain the complete set of pens as the number of pens increases? (In general, for N different pens, or outcomes, the expected number of boxes required is

---

214   TEACHING STATISTICS AND PROBABILITY

$N/1 + N/2 + N/3 + \ldots + N/N.$) The expected length can be plotted against the number of outcomes to help tell the story. The graph in figure 28.1 shows the values for four, eight, twelve, and sixteen outcomes. Each estimate, done with the help of a computer, is based on 100 trials. Some students may want to estimate expected values associated with fifteen or more equally likely outcomes.

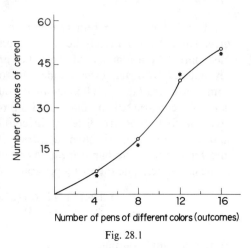

Fig. 28.1

Steps in using the Monte Carlo method

Model. Find an appropriate model for the problem situation. The six-sided die was suited for solving the six colored pens problem where 1 = pink, 2 = purple, 3 = auburn, 4 = rust, 5 = turquoise, and 6 = tawny.

Trial. A trial consists of rolling the die until a complete set of outcomes (here, a complete set of the six colored pens) is obtained. For example, these outcomes of throws of a die were obtained in a trial: 66433, 45332, 2561.

Trial	1	2	3	4	5	6	Boxes
1	/	/ /	/ / / /	/ /	/ /	/ / /	14

Result of trial. A trial has some numerical value associated with it. In our examples so far, it has been the number of rolls needed to obtain a complete set of outcomes—that is, the length of the trial. In other problems it may be the number of games won or some other result needed to solve the problem at hand. The outcome of the sample trial above is the number 14.

Repeat trials. The trials are repeated until an "appropriate number" has been obtained. The number of trials required for a specified level of accuracy can be determined mathematically (see, for example, [125] or [1]). For most problems encountered in schools, though, about a hundred trials will provide adequate accuracy.

Expected value of results of trials. Find the mean (average) value of all trials. For our example of six colored pens, this was found to be the expected number of rolls for complete sets of outcomes (based on only twenty trials):

$$\frac{288}{20} = 14.4$$

Probability

In the spirit of the previously used approach to finding expected values (of numbers of trials, of temperature, etc.), we now approach probability problems with a fresh point of view. Consider the following problem.

True-false test by guessing

Rick thinks he doesn't have to study for tests. He is willing to take his chances. Suppose he takes a true-false test in biology and doesn't know the answers to ten of the questions. What are his chances of getting seven or more of those ten questions correct by guessing?

This is a rather difficult question in probability. It is typically discussed when binomial probabilities are studied, and then it assumes a knowledge of the binomial expansion, among other things. But in our approach we "take" the true-false test lots of times, guessing at the answers, and keep a record of the number of correct answers obtained each time the test is "taken." A rather natural way to think of answering a true-false question by guessing is to flip a coin.

Steps for the Monte Carlo approach

The five steps for the Monte Carlo approach to probability are discussed below.

Model. We use a coin, where heads means "false" and tails means "true." In our approach, we provide the student with an "answer key" after the test has been taken. That is, the teacher tosses a coin to tell whether the answer to each item is true or false. In this way, students do not know until their tests are "scored" how many "correct answers" they obtained.

Trial. A trial consists of tossing the coin ten times, one for each question on the test, and then using the answer key (teacher's coin tosses) to determine the number of correct answers.

Successful trial. A successful trial occurs when seven or more correct answers are obtained.

Number of trials. A total of around a hundred trials should be obtained.

Probability (success). The probability of getting seven or more correct answers is estimated by the ratio

$$\frac{\text{Number of successful trials}}{\text{Total number of trials}}.$$

Figure 28.2 shows a sample answer sheet, a teacher's answer key (also obtained by tossing a coin), and a summary of the number of correct answers obtained from 100 trials. The summary shows that P (seven or more correct) is estimated as $21/100 = .21$.

Answer Sheet

Name _____*Brenda*_____

Questions	Teacher's key
1 T	1 T
2 F	2 F
3 F	3 F
4 F	4 T
5 T	5 T
6 F	6 T
7 T	7 F
8 T	8 F
9 T	9 F
10 T	10 T

N correct $= 5$

Distribution of Correct Answers (out of 10)

N	Tallies	Frequency
0		0
1	/	1
2		7
3		6
4		19
5		25
6		21
7		10 ⎫
8		8 ⎬ 7 or more correct
9		2 ⎭
10	/	1
		100

Fig. 28.2

Birth-month problem

What is the probability that in a group of four people chosen at random, at least two were born in the same month (not necessarily in the same year)?

Model. Use a twelve-sided die (one side for each month of the year), a table of twelve random digits, or the following BASIC program:

```
10  PRINT INT (12*RND(0)) + 1;
20  GO TO 10
30  END
```

Trial. A trial consists of rolling the die four times, once for each person in the group.

Successful trial. A successful trial is one in which a number is obtained more than once in the four rolls of the die—that is, at least two people have the same birth month. See the chart in figure 28.3 for a convenient method of keeping a record of trials.

Number of trials. Repeat for at least fifty trials.

Probability (success). The probability *(P)* that at least two people share the same birth month is estimated by the ratio

$$\frac{\text{Number of successes}}{\text{Number of trials}}.$$

Birth-Month Chart

	Jan.	Feb.	Mar.	Apr.	May	June	July	Aug.	Sept.	Oct.	Nov.	Dec.	
Trial	1	2	3	4	5	6	7	8	9	10	11	12	Success?
1				/				/		/	/		No
2			/ /	/		/							Yes

Fig. 28.3

This experiment was conducted for a total of 120 trials, and 47 successes were obtained; therefore

$$P \text{ (at least two people share birth month)} = \frac{47}{120} \approx .39.$$

The theoretical probability that in a group of four people picked at random at least two have the same birth month can be shown to be

$$1 - \left(\frac{12}{12}\right)\left(\frac{11}{12}\right)\left(\frac{10}{12}\right)\left(\frac{9}{12}\right) = 1 - \frac{990}{1728} \approx .43.$$

Some exploration can be done with this problem. How does the probability of a shared birth month vary with the number of persons in the group? Find probabilities for groups of two, six, eight, and ten persons and plot the results. Figure 28.4 shows the theoretical and experimental probabilities (based on 100 trials) for the birth-month problem.

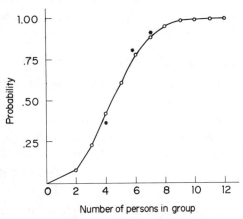

Experimental probabilities shown by *.

Fig. 28.4

It is left as an exercise for the reader to show how this method can be used to solve the birth*day* problem, which can be stated as follows:

What is the probability that in a group of, say, twenty-five people chosen at random, at least two have the same birthday?

Statistical Inference

This topic moves to a more complex level of thinking, but it, too, can be treated in a Monte Carlo fashion. Let's return to the problem of the true-false test.

A test of ESP

Rick says that he doesn't have to study for tests because he has ESP (extrasensory preception). He demonstrates his power to his doubting friends by taking a ten-item true-false test on topology, a subject he, knows nothing about. He gets seven of the items correct. What can we conclude about Rick's claim that he has ESP?

A way of thinking about this problem is first to recognize that it is possible to do well on a true-false test by pure luck, or guesswork. But we want to be able to say something about how often this might occur. How likely, therefore, is it that a person will get a score of seven or more correct on a ten-question true-false test by guesswork? Information about the likelihood of this occurring is already available to us from our earlier problem about Rick. Notice that some very high scores (but also some very low scores) were obtained by chance—by tossing a coin. We saw that the probability of getting seven or more correct answers by guessing is $21/100 = .21$. This is not really an unusual event, since it happened 21 out of 100 times. So, we are not willing, based on this evidence, to conclude that Rick has ESP.

Steps for doing statistical inference

The steps for doing statistical inference using a Monte Carlo approach are described below.

Model. A model, such as a coin or die, is needed to produce outcomes having a specified probability. In the ESP problem we use a coin, since we want the outcomes to represent guessing the answers to a true-false test.

Heads—False
Tails—True

Trial. Toss a coin ten times (one for each question on the test).

Outcome of a trial. The value of the outcome of a trial is expressed in terms of the "statistic of interest," that is, in terms of the number of correct answers (or whatever is specified in the problem).

Number of repeated trials. About a hundred trials should be obtained.

Probability of obtaining the statistic of interest. In this step we determine, on the basis of our Monte Carlo experiment, the probability of obtaining a value as great or greater than the one reported in the problem. In our example, we estimated the probability of obtaining seven or more correct answers as .21, based on 100 trials.

In Monte Carlo methods for statistical inference, we add a sixth step— the decision, or inference, that can be made on the basis of the experiment conducted.

Decision. We have estimated the probability of obtaining seven or more correct answers as .21. Since this is a fairly common event, we conclude that the model of the coin is appropriate for explaining Rick's score. That is, on the basis of our experiment, we conclude that he was, after all, only guessing and does *not* have ESP.

Why Use Monte Carlo Methods?

Many principles of good teaching are naturally embodied in Monte Carlo methods. First, there is activity. Students are required to do something. They toss coins or roll dice and contribute their results to the solution obtained by the class as a whole. Also, the method is concrete. Students can see the coins, toss them, and say, "Look at my result. I got two heads and three tails this time. That stands for two girls and three boys." When exploring the expected-value problem of the number of boxes of cereal needed to obtain a complete set of pens, the student has actually rolled a die and seen that it took, at times, as few as eight or ten rolls and as many as twenty-five or more rolls to obtain a complete set of outcomes. Then, when the average number of rolls of the die is calculated, the student can compare the result to his or her own experience of rolling a die.

Concepts in probability and statistics can be meaningfully taught through the use of Monte Carlo methods to students who have rather limited mathematical backgrounds. Many of these concepts are conventionally taught in advanced probability and statistics courses in the secondary school or in first- or second-year college mathematics courses. The use of Monte Carlo methods opens up access to a large variety of believable, real-world problems. The problems presented here have been selected to demonstrate something of the range of topics that can be addressed.

Of course, Monte Carlo methods do not answer all the difficulties in teaching probability and statistics. They require careful planning on the part of the teacher and, most important, careful attention to students' responses. When using Monte Carlo methods, one must be ready for the unexpected. Unusual outcomes do occur. There is also the danger that students will be led through the problems too quickly, not being allowed the opportunity to think carefully about what they have done and to move with confidence from one stage of thinking to another. This danger is great, perhaps, because the concrete, active, or even gamelike nature of Monte Carlo methods can lead one to lose track of the key and often profound concepts that are emerging. However, as one teacher remarked after a two-week Monte Carlo workshop, "I came here thinking this was just a gimmick—a sugar coating for a bitter mathematical pill. Now I see that it is a mathematical method with solid substance behind it. My students can profit a great deal from learning Monte Carlo methods. I feel very happy about that."

VIII. Using Computers

29. Solving Probability Problems through Computer Simulation

William Inhelder

COMPUTER simulation can reveal empirical solutions to many probability problems. Prior to the general availability of computers, mathematics teachers had recourse to two approaches in teaching students to solve probabilistic problems:

1. Physically carry out the probability experiment a significant number of times.
2. Solve the problem in an a priori fashion, using the laws of combinatorics and probability theory.

In the first instance, the physical and time constraints severely limited the nature and scope of the problems that could be investigated. The second approach, although much more efficient and mathematically satisfying, often required a level of mathematical sophistication beyond that of the student. Again, this necessitated limiting both the nature and the scope of the probability problem.

Easier access to computers and the increasing number of students with programming ability has changed all this. Most probabilistic problems can be readily simulated and the results of repeated simulation recorded to discover, or at least conjecture, what rule or principle is operating. Once the principle is discovered, it is expected that the student will be motivated to learn the necessary laws of combinatorics and probability to prove it true.

These somewhat unusual classroom probability problems are suitable for computer simulation:

1. The duration of a penny-matching game
2. The number of matches in two sets of N objects and their associated probabilities (a generalization of the hatcheck problem)

220

3. The probabilities involved with a deck of ESP cards in carrying out an experiment in extrasensory perception (ESP) in the fashion of J. B. Rhine at his Parapsychology Laboratory at Duke University

4. The probability of unacceptable queue lengths at checkout counters in a supermarket with a constant rate of random customer arrival and a normally distributed checkout time interval

5. The probability of randomly written four-letter English words

Only the first example can be carried out on a programmable calculator. If the calculator does not have a random number function key, then an algorithm incorporating a simple random number generator will have to be included in the program. Most manufacturers supply such algorithms. Because of the storage constraints and the number of iterations needed, the remaining examples can be handled adequately on microprocessors or minicomputers. If execution time is important, then a larger computer system is advised. (Programs for these five problems were written in BASIC and run on a Burroughs 6700 computer system. The program for the third problem is given in fig. 29.1. The rest of this article treats the third problem.)

For a simulation to be useful, it must accurately reflect the real situation. It is important, therefore, to identify and make explicit those constraints that appear to be relevant in the real situation so they can be programmed into the simulation. Problems occur when it is uncertain whether specific constraints are really relevant or not. In addition, if the expectation in the real situation is not fairly well defined, it is very difficult to test the validity of the simulation. Fortunately, most probability problems are fairly well defined. However, in the last example, since explicit, well-defined rules for writing a four-letter English word do not exist, we must develop and rely on statistically derived rules.

In order to make more explicit some of the problems with the simulation of a probabilistic event and the interpretation of the results, the problem of the ESP card simulation experiment will be presented.

To determine whether an individual possesses some degree of extrasensory perception, the following experiment is carried out with two persons, the "transmitter" and the "receiver." A deck of twenty-five ESP cards (also called Zener cards after an associate of Rhine's at Duke University) contains five different symbols, each depicted on five cards. The deck is shuffled, and the transmitter, turning over one card at a time, sends a "mental image" to a receiver, who notes each of the symbols she or he "receives." The total number of matches is then claimed to be a measure of the receiver's ESP. Of course, an evaluation of high ESP is strengthened by repeated trials of the experiment where the subject's score consistently exceeds the average. The probability given by the manufac-

Basic Program for ESP Experiment
(Dependent Strategy—Assuming No ESP)

```
 10 PRINT "THIS PROGRAM SIMULATES THE RESULTS OF A STANDARD ESP"
 20 PRINT "CARD EXPERIMENT. A 25 CARD DECK CONTAINS 5 DIFFERENT"
 30 PRINT "SYMBOLS OF 5 CARDS EACH. THE DECK IS SHUFFLED AND THE"
 40 PRINT "TRANSMITTER, TURNING OVER ONE CARD AT A TIME, SENDS A"
 50 PRINT "MENTAL IMAGE TO A RECEIVER WHO NOTES EACH OF THE
       SYMBOLS"
 60 PRINT "HE RECEIVES. THE PROGRAM PRINTS OUT THE FREQUENCY"
 70 PRINT "DISTRIBUTION OF THE NUMBER OF MATCHES FOR 1000
       REPETITIONS"
 80 PRINT "OF THE EXPERIMENT."
100 DIM A(25),B(25),C(25)
110 X1 = 100*(BCL-INT(BCL))
120 X2 = X1+3
140 FOR I = 0 TO 25
150 C(I) = 0
160 NEXT I
165 FOR K = 1 TO 1000
170 FOR I = 1 TO 25
180 A(I) = B(I) = I
190 NEXT I
200 FOR J = 1 TO 25
210 R1 = INT(25*RND(X1)+1)
220 H = A(J)
230 A(J) = A(R1)
240 A(R1) = H
250 NEXT J
260 FOR J = 1 TO 25
270 R2 = INT(25*RND(X2)+1)
280 H = B(J)
290 B(J) = B(R2)
300 B(R2) = H
310 NEXT J
312 FOR I = 1 TO 25
314 A(I) = A(I)MOD 5
316 B(I) = B(I)MOD 5
318 NEXT I
320 S = 0
330 FOR I = 1 TO 25
340 IF A(I) = B(I) THEN S = S+1
350 NEXT I
360 C(S) = C(S)+1
370 NEXT K
375 PRINT "NUMBER OF MATCHES          FREQUENCY"
380 FOR I = 0 TO 25
385 IF C(I)=0 THEN 410
400 PRINT TAB(7);I;TAB(25);C(I)
410 NEXT I
420 END
```

Annotations (right of code):

- Lines 110–120: These statements generate different time-dependent arguments for the RND function.
- Lines 140–160: Initializes variables for tallying frequencies.
- Lines 200–250: Shuffles ESP card deck.
- Lines 260–310: Scrambles responses from 1 to 25.
- Lines 312–318: Changes shuffled deck and scrambled responses into 5 different symbols of 5 cards or responses each.
- Lines 330–360: Tallies the number of matches in both sets.

Fig. 29.1

turer of the deck of cards seems to indicate that the binomial probability distribution was used with $p = 1/5$ and $q = 4/5$. However, this distribution is to be used only when the repeated events are independent. If each card were returned to the deck after its image was transmitted and then reshuffled before the next card was drawn, then the events would indeed be independent and the binomial probabilities would hold. In the actual experiment, since the receiver knows that each symbol can appear no more than five times, each response would seem to depend on previous responses. Might not the probabilities under these conditions differ significantly from those predicted by the binomial distribution?

Because an a priori analysis of dependent probabilities seemed overwhelming, a computer simulation of the experiment was attempted. In this simulation of the strategy of the receiver, assuming no ESP is involved, it was felt that the subject would certainly not consciously put down more than five of any symbol and would probably subconsciously tend to assign different probability weights to symbols in some rough proportion to the degree to which these symbols had already occurred. This dependent strategy and a second independent random strategy were simulated and investigated for 10 000 trials each. Note that in the second strategy it is possible to have varying degrees of unequal occurrences of the symbols.

Objectives of the simulation

1. Determine the expected mean number of matches in the simulated experiment. How does this compare with the expected mean of five matches predicted by the binomial distribution?

2. Develop a table of probabilities for the various number of matches in the simulated experiment. How does this compare with the probability table predicted by the binomial distribution?

Results and interpretation

In table 29.1 the experiment was simulated 10 000 times. Probabilities of the various numbers of matches for dependent and independent strategies are recorded. The binomial probabilities were developed using $P(r$ matches$) = {}_nC_r p^r q^{n-r}$. Thus $P($exactly six matches$) = {}_{25}C_6(.2)^6(.8)^{19}$. It appears that the receiver's strategy is irrelevant, since the receiver does

TABLE 29.1

COMPARISON BETWEEN BINOMIAL PROBABILITIES AND
PROBABILITIES OF SIMULATED ESP EXPERIMENT

Number of Matches	SIMULATED EXPERIMENTAL PROBABILITY IN PERCENT		Binomial Probability in Percent
	Dependent	Independent	
0	0.38	0.38	0.38
1	2.53	2.62	2.4
2	7.31	7.13	7.1
3	13.59	13.13	13.6
4	18.69	19.19	18.7
5	19.03	20.09	19.6
6	15.71	15.87	16.3
7	11.03	11.13	11.1
8	6.37	5.95	6.2
9	3.20	2.82	2.9
10	1.42	1.22	1.2
11	0.47	0.36	0.32
12	0.24	0.07	0.12
13	0.03	0.04	0.002
	$\overline{X} = 5.0170$	$\overline{X} = 4.9712$	$\mu = 5.0$
	$\tilde{\sigma} = 2.0573$	$\tilde{\sigma} = 1.9897$	$\sigma = 2.0$

not know the order of the cards that the transmitter turns up (assuming no ESP). Hence, for the purpose of matching cards from the two sets, the receiver's response may be assumed to be independent of the cards turned up by the transmitter. This accounts, in part, for the similarity of the distributions with the binomial distribution. However, does the fact that the receiver knows there can be no more than five of each type change either the mean or the nature of the distribution? The answer appears to be no.

The crucial aspect in analyzing the experiment is that a receiver's responses, with no ESP, are actually independent of the transmitter regardless of whether the receiver uses a dependent or an independent random strategy. Clearly a strategy of selecting twenty-five identical symbols, although producing a mean of five and a standard deviation of zero, could not result in a binomial distribution. Furthermore, a strategy of selecting twelve of one symbol and thirteen of another would yield a distribution of zero to ten matches with a mean of five. Because of the relative infrequency of matches above ten, the distribution resembles the binomial distribution but is a poor fit. Finally a strategy of eight of one symbol, eight of another, and nine of yet another would create a distribution of zero to fifteen matches, which is fairly close to the binomial distribution. Thus the larger the number of different cards, the closer the distribution is approximated by the binomial distribution. This conjecture was verified by additional trials of the ESP simulation using the strategies above. That there are not five of each is less important than the fact that five different cards are represented in the set. All this would appear to explain why the receiver's different strategies produced the same results and why the binomial distribution serves as a good approximation for the probabilities in the actual ESP experiment.

Computer simulation can help students discover and develop probabilistic truths in a realistic problem. Without computer simulation, gathering sufficient experimental data to investigate this problem would be so time-consuming that it would not be feasible for the classroom. In addition, the numerous ways of confounding the results are avoided through simulation. A comprehensive presentation of the difficulties with Rhine's ESP experiments and results can be found in [15].

30. In All Probability, a Microcomputer

Howard M. Kellogg

Probability is an especially good area for using a school's microcomputer creatively because the dynamic nature of a stochastic event fits well with the computer's interactive and display capabilities. Further, the computer in a teaching setting can be an interactive device, encouraging students to think, analyze, make logical decisions, come to understand a situation with mathematical content, and, in understanding the situation, come to understand the mathematics as well.

Often we say to our students, "Here is a problem; solve it." With microcomputers we have, more than before, the opportunity to say, "Here is a situation. What can you say about it? What problems does it suggest? How would you solve them?"

This article will present several probability problems for the microcomputer that go well beyond traditional computation. Simulation will play a central role in each. In each example, various uses can be made of the simulation, leading to a deeper understanding of the situation it represents and, thus, of the mathematics involved. Since a program written in BASIC is included for each simulation, students can be encouraged to modify those programs to vary the simulation—for instance, they could run it a thousand times to collect data on some unknown parameter. Suggestions are given for each example on the kinds of activities that might be undertaken, but it would be desirable to let suggestions come from the students themselves. These may surprise you!

The best simulations from a teaching point of view often use some sort of graphics display, since this helps students quickly form a clear concept of the situation represented. In probability, especially, the situation is usually undergoing some change, and these changes can be dynamically represented in a graphics display. The examples presented here assume the use of an Apple II microcomputer with a color television attached. The Apple II system has a fairly advanced graphics capability and is being used by many schools. The programming language is the Applesoft II version of BASIC.

The Tontine

Our first example involves a *tontine,* a primitive (and now illegal) form of life insurance. In a tontine, each member of a group contributes a cer-

The author wishes to thank Bill McNamara, owner of Computerland, Paramus, New Jersey, and his staff, for making available computer facilities and for providing technical assistance.

225

tain amount of money and agrees that the last survivor of the group will win the whole sum. (Can you guess why tontines are now illegal?) It is possible to develop a formula for the probability that a given participant will win the tontine, but the derivation is tedious and not terribly instructive. The approach taken here is quite different: we shall simulate the operation of a tontine. We can then run the simulation a large number of times to obtain statistical estimates for such parameters as the probability that a particular person will win or the waiting time for a winner.

To simulate the operation of a tontine, we shall create a list of names and ages to represent the participants, display the list on a CRT or graphics terminal, and then remove names from the list by a random process to simulate the gradual dying off of the group. When one name is left, the process stops, and that person is, of course, the winner.

A table of mortality factors will be needed to simulate the dying-off process. Each mortality factor, denoted q_x, is the probability that a person of age x will die within one year. Except for advanced ages, these probabilities will be rather small; for example, one estimate for q_{60} used by insurance companies is 0.013 119. It is usually assumed that for some advanced age x, such as 110, q_x will be 1, that is, death within one year at this age is certain. This age is denoted ω (omega), the end of the mortality table. A typical table of mortality factors appears in Jordan [62].

How shall we set up the dying-off process for our list of names? Our program will take each name in turn, calculate that person's current age, then use the mortality factor for that age (the probability of dying within one year) and a random number generator to simulate the occurrence or nonoccurrence of the event "dies within the year." If the event occurs, then that name will be removed from the list. Each time the program passes through the list represents the passage of one year. Our program will also display the year currently represented.

Most versions of BASIC include a random number generator, essential in this and subsequent programs. In the version used here, one obtains a random number between 0 and 1 with the statement RND(1). Since in theory each number in this interval has an equal probability of being chosen, the probability of that number being between 0 and 1/2 is 1/2, between 0 and 1/4, 1/4, and so forth, and between 0 and p $(0 < p < 1)$, p. Thus to simulate an event E with probability p, we get a random number between 0 and 1 and test whether it is between 0 and p. If it is, E has occurred; if not, E has not occurred. The short routine of figure 30.1 does this, setting J = 0 if the event does not occur and J = 1 if it does. This routine can be used in many simulation applications.

To avoid having to key in the mortality factors one at a time each time the simulation program is run, it is convenient to create an array of mortality factors in a separate program and then store that array on tape. The program listed in figure 30.2 will do this.

```
1000 REM GIVEN P, SIMULATES AN EVENT WITH PROBABILITY
     P:RETURNS 1 IF EVENT OCCURS, 0 IF IT DOES NOT
1010 J = 0
1010 R = RND (1)
1030 IF R < P THEN J = 1
1040 RETURN
```

Fig. 30.1. Routine to simulate an event with probability p

```
10 PRINT "INPUT OMEGA";:INPUT OMEGA
20 DIM AGE(OMEGA)
30 PRINT "INPUT AGE OF FIRST Q";:INPUT QIX
40 PRINT "INPUT MORTALITY FACTOR 5 FROM AGE ";QIX;" TO
   AGE"; OMEGA: PRINT
50 FOR I = QIX TO OMEGA
60 PRINT "AGE";I;"QZ=";:INPUT AGE(I)
70 NEXT I
80 PRINT "READY TAPE, PRESS ANY KEY WHEN SET"
90 GET A$
100 STORE AGE
110 PRINT "DONE"
120 STOP
999 END
```

Fig. 30.2. Program to put mortality factors on tape

If the array is to be stored on disk, the following lines should be inserted in the program:

```
80 D$=CHR$(4)
82 PRINT D$; "OPEN MORTALITY"
84 PRINT D$; "DELETE MORTALITY"
86 PRINT D$; "OPEN MORTALITY"
88 PRINT D$; "WRITE MORTALITY"
90 FOR J=1 TO OMEGA
95 PRINT AGE(J)
100 NEXT J
105 PRINT D$; "CLOSE MORTALITY"
```

The program that simulates the operation of the tontine is shown in figure 30.3. In line 150, the array of mortality factors previously created is read in from tape. Other entries are keyed in: the names and ages of the participants, omega, and the starting year of the tontine.

```
 90 PRINT "OMEGA=";  INPUT OMEGA
100 PRINT "INPUT NUMBER OF PERSONS IN TONTINE"; INPUT N
110 DIM NAME$(N)
111 DIM AGE(N)
112 DIM ALIVE(N)
113 DIM QX(OMEGA)
120 FOR I = 1 TO N:ALIVE(I) = 1: NEXT I
130 PRINT "FOR EACH PERSON IN GROUP INPUT NAME AND
    PRESENT AGE"
```

```
140 FOR I = 1 TO N: PRINT "N= "; I;" "
142 INPUT NAME$(I)
144 PRINT "   ";: INPUT AGE(I):NEXT I
150 RECALL QX
160 PRINT "STARTING YEAR="; INPUT SYEAR
170 CYEAR = SYEAR: JX = 0: SUM = N
180 REM LOOP STARTS HERE
185 HOME
190 PRINT "STARTING YEAR OF TONTINE=";SYEAR
195 CYEAR = SYEAR + JX
200 PRINT "CURRENT YEAR="; CYEAR
210 PRINT:PRINT
220 FOR I = 1 to N
230 IF ALIVE(I) = 0 THEN GOTO 300
240 M = AGE(I): P = QX(M)
250 GOSUB 1000
260 IF J=0 THEN PRINT NAME$(I): HTAB(30): PRINT AGE(I):
    GOTO 290
270 ALIVE(I)=0: SUM=SUM-1
280 IF SUM=1 THEN GO TO 500
290 AGE(I)=AGE(I)+1
300 NEXT I
310 JX=JX+1
320 GOTO 185
500 REM END SEGMENT
510 HOME
520 PRINT "STARTING YEAR OF TONTINE="; SYEAR
530 CYEAR=SYEAR+JX+1
540 PRINT "ENDING YEAR OF TONTINE="; CYEAR
550 PRINT:PRINT:PRINT:HTAB(136):PRINT "WINNER!"
560 PRINT:PRINT:PRINT:PRINT NAME$(I): HTAB(30): PRINT AGE(I)
570 STOP
1000 REM GIVEN P, SIMULATES AN EVENT WITH PROBABILITY P:
     RETURNS 1 IF EVENT OCCURS, 0 IF IT DOES NOT
1010 J=0
1020 R = RND (1)
1030 IF R < P THEN J = 1
1040 RETURN
```

Fig. 30.3. Program to simulate a tontine

If disk is to be used, replace line 150 with the following:

```
150 D$=CHR$($)
152 PRINT D$; "OPEN MORTALITY"
153 PRINT D$; "READ MORTALITY"
154 FOR J=1 TO OMEGA
155 INPUT QZ(J)
156 NEXT J
157 PRINT D$; "CLOSE MORTALITY"
```

The running of this program is rather striking. The names all appear on the screen at the start. At the top, the years flash by quickly, but for a while nothing else is likely to happen because the probability of dying at the lower ages is small. Then, suddenly, a name disappears; the list shortens. After an interval, another name disappears, then another, then faster, until suddenly only one name remains.

This program provides students with the raw material that I call a mathematical situation. Using the program, students can investigate in an empirical way such questions as, What is the probability that the youngest person wins? The oldest? The person with the median age? How does the spread of ages affect the probabilities? What is the average length of time until a winner emerges? To investigate some of these questions, students will want to modify the program in figure 30.3 so that repeated runs take place automatically and so that records can be kept on such data as the number of times a particular person wins or the wait for a winner.

This program does not require complete graphics capability, only a CRT or television output. Nevertheless, the running program certainly has a dynamic display that puts it squarely in the realm of graphics instructional programs.

Monte Carlo Estimation of Area

Estimating the area of a region is commonly given in BASIC texts to illustrate the use of the random number function. Adding a graphics capability and making this technique visual will greatly increase its impact. Students' intuitive grasp of the concept *random* will be enhanced.

Suppose we want to find the area of the shaded region shown in figure 30.4. If a random point is chosen within the rectangular border, the probability that the point will lie within the shaded region is proportional to the area of that region. For example, if the shaded region is one-fifth of the area enclosed by the rectangle, then the probability that the random point will be inside the shaded region, given that it is within the rectangle, is one-fifth. If, say, 1000 points were randomly located within the

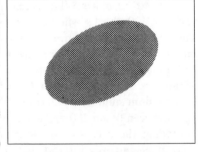

Fig. 30.4. A region of unknown area

rectangle, we should expect about 200 of them to be inside the shaded region.

If the area of the shaded region is unknown, we can estimate it by randomly placing a large number of points within the rectangle, observing the proportion, $\hat{p}$, of them falling within the shaded region, then multiplying $\hat{p}$ by the area of the rectangle. The result is an estimate for the area of the shaded region. By the way, $\hat{p}$ is an estimate for the probability, p, that a randomly chosen point will lie within the shaded region.

All that has been described so far in this example can be done by any computer with a random number function, including those without a graphics display. How much more convincing this process becomes when one sees the randomly chosen points appear, one at a time, on a screen showing the region of unknown area! A program that does this is presented in figure 30.5.

```
10  S = 0
20  HGR
30  HCOLOR= 1
40  FOR Y = 40 TO 120
50  HPLOT 70,Y TO 210, Y
60  NEXT Y
70  PRINT "INPUT NUMBER OF POINTS TO BE PLOTTED";:
    INPUT N
80  HCOLOR=5
90  FOR J = 1 TO N
100 K = INT (280 * RND (1)):L = INT (160 * RND (1))
110 HPLOT K,L
120 IF K > = 70 AND K < = 210 AND L > = 40 AND L
    < = 120 THEN S = S + 1
130 NEXT J
140 R = S / N
150 PRINT "RATIO OF HITS TO TRIALS ="; R
160 PRINT "# TRIALS=";N, "#HITS =";S
170 STOP
999 END
```

Fig. 30.5. Program to estimate area by Monte Carlo method

Lines 40–60 plot on the screen the region whose area is to be estimated, and line 120 checks whether the random point (K,L) is within the region. The region in this example is a rectangle measuring (in screen coordinate units) 140 by 80. The entire screen in the Apple II system measures 280 by 160 units, with a space for text at the bottom. Since $RND(1)$ always gives a random number between 0 and 1, we multiply in statement 100 the first random number by 280 and the second by 160 to obtain an x coordinate between 0 and 280 and a y coordinate between 0 and 160.

Finding the area of a rectangular region in this way may not seem to hold much intrinsic interest, but it can serve to introduce the method to a class. Further, since the area of the region is already known, they can see how close to the true area their estimates are. One further experiment along these lines would be to perform a series of, say, twenty trials of 100 points each and plot the values found, that is, the estimated areas, to observe the clustering effect. This clustering could then be compared with the clustering obtained with twenty trials of 1000 points, which would be a much tighter clustering. These comparisons will show the effect on the distribution of a sample mean of increasing the sample size, an important concept in elementary statistics.

It is easy to substitute other regions for the rectangular region in the program of figure 30.5. Lines 40–60 must be changed to display the new region, and line 120 must be changed so that the displayed region is the one for which area is estimated. Be sure the region of lines 40–60 and the one of line 120 are the same! For example, if you wished to estimate the area of a circle of radius 50 with its center at the center of the screen, the new lines would be

```
40  FOR Y=30 TO 130
50  HPLOT -SQR(2500-(Y-80) * (Y-80))+140, Y TO SQR
      (2500-(Y-80) * (Y-80))+140, Y
60  NEXT Y
 .
 .
 .
120 IF(K-140) * (K-140)+(L-80) * (L-80)<=2500 THEN S=S+1
```

If we call the estimated area $\widehat{A}$, since the area of the circle is $A = \pi r^2$, π can be estimated by taking $\pi \approx \widehat{A}/r^2$, or $\pi \approx \widehat{A}/2500$. It will surprise students that out of this simple process of using random numbers comes a way of estimating π.

Another use that can be made of this method of estimating areas is to take an elliptical region, such as $400x^2 + 225y^2 \leq 1$ and use the Monte Carlo program to get a good estimate for its area. This must be transformed to $400(X - 140)^2 + 225(Y - 80)^2 \leq 1$ to center it on the screen of the Apple II.) Then vary first one of the parameters and then the other, creating a table showing area versus parameters. How does the area change as the parameters change? Can you guess the formula for the area of an ellipse? What about the area inside the parabola $y = x^2$ under the line $y = a$?

Buffon's Needle

A famous problem, Buffon's needle problem, goes like this: If a needle of length L is allowed to fall randomly on a surface on which is ruled a series of parallel lines a distance D apart, what is the probability that the needle will touch or cut across one of the lines? A little calculus shows that this probability is given by $p = 2L/\pi D$. An excellent discussion of this problem, including a derivation of the formula, is given in a *Mathematics Teacher* article by Schroeder [105]. An empirical estimate for p can be found by actually dropping a needle from a height of five feet or so onto a large sheet of appropriately ruled paper taped to the floor. If this observed estimate for p is (as usual) denoted $\hat{p}$, then an estimate for π is $2L/\hat{p}D$.

We shall simulate this experiment using a microcomputer with a graphics display. The program is given in figure 30.6.

This program is organized as a main line (lines 100–240) and three subroutines (starting at lines 1000, 2000, and 3000). After the main line gets as input the distance D between parallel lines and the length L of the needle (in screen coordinate units), the first subroutine (lines 1000–1120) sets up the display. The second subroutine (lines 2000–2080) gets a random initial point (x_0, y_0) and a random angle θ from that point, calculates the other endpoint (x_1, y_1) of the segment representing the needle, and then plots that segment. The third subroutine checks whether that segment intersects any of the parallel lines. The number of trials and the ratio of hits to trials is constantly displayed at the bottom of the screen. Line 230 checks whether any key on the keyboard has been pressed; if so, then it waits before plotting the next segment, allowing the viewer time to write down an intermediate value of the ratio before continuing. Then the program goes back to redo the display, drop the needle again, and so on.

One use of this program is to let students find the formula empirically by varying D and L and using a large number of trials (say 500) for each D and L chosen. This can be done much more quickly and easily by simulation, of course, than by actually dropping needles. You may (or may not) wish to give the hint that the formula contains π. Also, you might ask whether p would be expected to increase or decrease if (1) L is increased and (2) D is increased. Encourage systematic rather than haphazard exploration, start with "reasonable" values of L and D. What happens if D doubles? If you give students the formula (or better yet, if they've obtained it as described above), have them use it to obtain estimates for π. How quickly does it appear to converge?

Conclusion

I have attempted to show by example that in the teaching of probability in the secondary school microcomputers have a possible role that goes considerably beyond their use as supercalculators. Simulations by means of a microcomputer can expose students to situations that are rich in mathematical content; the students themselves decide how to analyze each situation, sometimes even deciding what the problems will be and how they will be solved. This really is interactive learning. What's more, working with these simulations is a lot of fun!

Although probability is an especially rich field for finding mathematical situations that can be simulated with a microcomputer, don't forget that similar activities can be drawn from other areas as well. Finally, remember that plenty of computer applications of the more traditional kind can be found in the pages of the *Mathematics Teacher* and elsewhere, many dating from before the era of the microcomputer. Since the language used for these was usually BASIC, many of them can be put on a microcomputer with only minor adaptations. In all probability, however you look at it, a microcomputer should be an important piece of hardware in your professional tool kit.

```
100   INPUT M,L
110   S = 0:T = 0
120   GOSUB 1000: REM DISPLAY
130   HCOLOR= 5
140   GOSUB 2000: REM PLOT SEGMENT
145   FOR I = 1 TO 500: NEXT I
150   GOSUB 3000: REM HIT?
160   IF J = 1 THEN S = S + 1
170   T = T + 1
180   VTAB 21
190   R = S/T
200   PRINT "NUMBER OF TRIALS=";T
210   PRINT "RATIO OF HITS TO TOTAL =";R
220   IF PEEK (-16384) < = 127 THEN 120
225   POKE - 16368,0
230   IF PEEK (- 16384) < = 127 THEN 230
235   POKE - 16368,0
240   GOTO 120
1000  REM DISPLAY BORDER AND LINES
1005  HGR
1010  N = INT (279 / M) + 1
1020  A = (279 - (N - 1) * M) / 2
1030  HCOLOR= 6
1040  HPLOT A,1 TO A + (N - 1) * M,1
1050  HPLOT A, 159 TO A + (N - 1) * M, 159
1060  HPLOT A, 1 TO A, 159
1070  HPLOT A + (N - 1) * M,1 TO A + (N - 1) * M, 159
1080  HCOLOR = 3
1090  FOR I = 1 TO N - 1
1100  HPLOT A + I * M,2 TO A + I * M, 158
1110  NEXT I
1120  RETURN
2000  REM GET RANDOM POINT AND ANGLE, CALCULATE 2ND
      POINT, PLOT THEM
2010  X0 = (N - 3) * M * RND (1) + 2 * M + A
2020  Y0 = 159 * RND (1)
2030  RHO = 6.283184 * RND (1)
2040  X1 = X0 + L * COS (RHO)
2050  Y1 = Y0 + L * SIN (RHO)
2060  IF X1 < 1 OR X1> 279 or Y1 < 0 or Y1> 159 THEN 2030
2070  HPLOT X0, Y0 TO X1, Y1
2080  RETURN
3000  REM CHECK IF HIT:RETURNS J=1 IF YES, J=0 IF NO
3010  J = 0
3020  FOR I = 1 TO N
3030  IF (X0 < (I-1) * M + A) AND (X1 < (I-1) * M
      + A) THEN GOTO 3060
3040  IF (XO > (I - 1) * M + A) AND (X1 > (I - 1) * M + A) THEN
      GOTO 3060
3050  J = 1
3060  NEXT I
3070  RETURN
9999  END
```

Fig. 30.6. Program to simulate Buffon's needle problem

Bibliography

Compiled by

Stuart A. Choate

1. Atkinson, David T., C. Shevokas, and K. J. Travers. *Probability and Statistics.* Champaign, Ill.: Stipes Publishing Co., 1976.
2. Baratta-Lorton, Mary. *Mathematics Their Way,* pp. 140–63. Menlo Park, Calif.: Addison-Wesley Publishing Co., 1976.
3. Barton, R. F. *A Primer on Simulation and Gaming.* Englewood Cliffs, N.J.: Prentice-Hall, 1970.
4. Battacharrya, G. K., and R. A. Johnson. *Statistical Concepts and Methods.* New York: John Wiley & Sons, 1977.
5. Bell, E. T. *The Development of Mathematics.* New York: McGraw-Hill Book Co., 1940.
6. Bickel, Peter J., Eugene A. Hammel, and William J. O'Connell. "Sex Bias in Graduate Admissions: Data from Berkeley." In *Statistics and Public Policy,* edited by William B. Fairley and Frederick M. Mosteller. Reading, Mass.: Addison-Wesley Publishing Co., 1977.
7. Billstein, Rick. "A Fun Way to Introduce Probability." *Arithmetic Teacher* 24 (January 1977): 39–42.
8. Blythe, Colin R. "Some Probability Paradoxes in Choice from among Random Alternatives." *Journal of the American Statistical Association* 67 (June 1972): 366–81.
9. Bright, G. W., J. G. Harvey, and M. M. Wheeler. "Achievement Grouping with Mathematics Concept and Skill Games." *Journal of Educational Research* 73 (1980): 265–69.
10. Bruni, James V., and Helene Silverman. "Graphing as a Communication Skill." *Arithmetic Teacher* 22 (May 1975): 354–66.
11. Bryson, Maurice. "The Literary Digest Poll: Making of a Statistical Myth." *American Statistician,* 30 November 1976, pp. 184–85.
12. Byers, Joseph W. "Lunch Money—Nuisance or Opportunity?" *Arithmetic Teacher* 18 (January 1971): 57–58.
13. Castellano, Janet, and Matthew Scaffa. *Unimath—Mathematics Activities for the Primary Grades Using Interlocking Cubes,* pp. 12, 14, 15. Palo Alto, Calif.: Creative Publications, 1974.
14. Choate, Stuart A. "Activities in Applying Probability Ideas." *Arithmetic Teacher* 26 (February 1979): 40–42.
15. Christopher, Melbourne. *ESP, Seers & Psychics,* pp. 8–38. New York: Thomas Y. Crowell, 1970.
16. Clarke, R. D. "An Application of the Poisson Distribution," *Journal of the Institute of Actuaries* 72 (1946): 481.
17. Clark, Malcolm, and Andrew MacNeil. "Odd Couples and Missing Cars." *Mathematics Spectrum* 9 (1976–77): 42–46.
18. Cohen, J. *Chance, Skill, and Luck: The Psychology of Guessing and Gambling.* Baltimore, Md.: Penguin Books, 1960.
19. _____. "Subjective Probability," *Scientific American* 197 (November 1957); 129–38.
20. Cohen, J., and M. Hansel. *Risk and Gambling.* New York: Philosophical Libraries, 1956.
21. Cohen, Morris R., and Ernest Nagel. *An Introduction to Logic and Scientific Method.* New York: Harcourt, Brace & Co., 1934.
22. Commission on Mathematics. *Program for College Preparatory Mathematics.* New York: College Entrance Examination Board, 1959.

23. Comprehensive School Mathematics Program. *Elements of Mathematics, Book 0, Intuitive Background.* St. Ann, Mo.: CEMREL, 1973.

24. ———. *Mathematics for the Intermediate Grades, Parts IV and V.* St. Louis, Mo.: CEMREL, 1978.

25. Conover, W. J. *Practical Nonparametric Statistics.* New York: John Wiley & Sons, 1971.

26. Consumers Union of the United States. *Consumer Reports.* Mt. Vernon, N.Y.

27. Cox, D. R., and David Hinckley. *Theoretical Statistics.* London: Chapman & Hall, 1974.

28. Davis, M. *Game Theory.* New York: Basic Books, 1970.

29. Edwards, W. "Conservatism in Human Information Processing." In *Formal Representations of Human Judgment,* edited by B. Kleinmutz. New York: John Wiley & Sons, 1968.

30. Ellis, Dorothy E. "Graphing." *Arithmetic Teacher* 25 (December 1977): 22.

31. Emerson, John D. "Introductory Statistics: A Contemporary Approach." *Mathematics Teacher* 70 (March 1977): 258–61.

32. Engel, Arthur. "Teaching Probabilities in Intermediate Grades." *International Journal of Mathematics Education in Science and Technology* 2 (February 1971): 243–94.

33. ———. "An Introduction to Probability," chap. 8. In *Elements of Mathematics, Book 0, Intuitive Background.* St. Louis: CEMREL, 1973.

34. Engel, Arthur, and Lennart Råde. "Topics in Probability and Statistics," chap. 12. In *Elements of Mathematics, Book 0, Intuitive Background.* St. Louis: CEMREL, 1973.

35. Enman, Virginia. "Probability in the Intermediate Grades." *Arithmetic Teacher* 26 (February 1979): 38–39.

36. Ericson, B. H., and T. A. Nosanchuk. *Understanding Data.* New York: McGraw-Hill Book Co., 1977.

37. Fairley, William B., and Frederick Mosteller. *Statistics and Public Policy.* Reading, Mass.: Addison-Wesley Publishing Co., 1977.

38. Feller, William. *An Introduction to Probability Theory and Its Applications.* Vol. 1, pp. 135–37. New York: John Wiley & Sons, 1957.

39. ———. *An Introduction to Probability Theory and Its Applications.* Vol. 2. New York: John Wiley & Sons, 1966.

40. Fischbein, E. *The Intuitive Sources of Probabilistic Thinking in Children.* Dordrecht, Holland: D. Reidel Publishing Co., 1975.

41. Fisz, M. *Probability Theory and Mathematical Statistics.* 3d ed. New York: John Wiley & Sons, 1963.

42. Fryer, M. J. *An Introduction to Linear Programming and Matrix Game Theory.* New York: John Wiley & Sons, 1978.

43. Galen, R. S., and S. R. Gambino. *Beyond Normality: The Predictive Value and Efficiency of Medical Diagnosis.* New York: John Wiley & Sons, 1975.

44. Gallup, George. "Opinion Polling in a Democracy." In *Statistics: A Guide to the Unknown,* edited by J. Tanur et al., pp. 146–52. San Francisco: Holden-Day, 1972.

45. Gardner, Martin. "The Paradox of the Nontransitive Dice and the Elusive Principle of Indifference." *Scientific American* 223 (December 1970): 110–14.

46. Gibbs, G. I., ed. *Handbook of Games and Simulation Exercises.* Beverly Hills, Calif.: Sage Publications, 1974.

47. Gilbert, Robert K. "Hey Mister! It's Upside Down!" *Arithmetic Teacher* 25 (December 1977): 18–19.

48. Gottlieb, B. S. *Programming with BASIC,* pp. 175–83. Schaum's Outline Series in Accounting. New York: McGraw-Hill Book Co., 1975.

49. Gray, Kenneth G., and Kenneth J. Travers. *The Monte Carlo Method.* Champaign, Ill.: Stipes Publishing Co., 1978.

50. Grillo, J. P. "A Comparison of Sorts." *Creative Computing* 2 (1979): 234–38.

51. Haack, Dennis G. *Statistical Literacy: A Guide to Interpretation.* Scituate, Mass.: Duxbury Press, 1978.

52. Hammersley, J. M., and D. C. Handscomb. *Monte Carlo Method.* London: Methuen & Co., 1964.

53. Hannigan, Irene. "Show It with a Graph!" *Teacher* 94 (January 1977): 79–80.

54. Hecht, James E. "Teaching Probabilistic Problem-Solving to Ninth Grade General Mathematics Students." Unpublished doctoral dissertation, University of Illinois, Urbana, 1980.

55. Hill, J. Ben, and Helen D. Hill. *Genetics and Human Heredity.* New York: McGraw-Hill Book Co., 1955.

56. Hirst, H. "Probability." *Mathematics Teaching,* no. 80, September 1977, pp. 6–8.

57. Hodges, J. L., David Krech, and R. S. Crutchfield. *Stat-Lab: An Empirical Introduction to Statistics.* New York: McGraw-Hill Book Co., 1975.

58. Huff, Darrell. *How to Lie with Statistics.* New York: W. W. Norton & Co., 1954.

59. Hyman, R. T. *Paper, Pencils and Pennies.* Englewood Cliffs, N.J.: Prentice-Hall, 1977.

60. Jacobs, Harold. *Mathematics: A Human Endeavor.* San Francisco: W. H. Freeman & Co., 1970.

61. Jones, Graham. "A Case for Probability." *Arithmetic Teacher* 26 (February 1979): 37, 57.

62. Jordan, C. W. *Life Contingencies.* Chicago: Society of Actuaries, 1952.

63. Kahneman, D., and A. Tversky. "Availability: A Heuristic for Judging Frequency and Probability." *Cognitive Psychology* 5 (May 1973): 207–32.

64. _____. "Judgment under Uncertainty: Heuristics and Biases." *Science* 185 (1974): 1124–31.

65. _____. "On the Psychology of Prediction." *Psychological Review* 80 (1973): 237–51.

66. _____. "Subjective Probability: A Judgment of Representativeness." *Cognitive Psychology* 3 (July 1972): 430–54.

67. Kaplan, Sandra Nina, JoAnn Kaplan, Sheila Madsen, and Bette Taylor. *Change for Children,* pp. 57–59. Santa Monica, Calif.: Goodyear Publishing Co., 1973.

68. Kaplan, Sandra Nina, Sheila Kunishima Madsen, and Bette Taylor Gould. *The Teacher's Choice: Ideas and Activities for Teaching Basic Skills,* pp. 180, 182, 189. Santa Monica, Calif.: Goodyear Publishing Co., 1978.

69. Keach, Everett T., Jr. "I Can Graph My Class." *Early Years* 9 (January 1979): 36–37.

70. Kemeny, J. G., J. L. Snell, and G. L. Thompson. *Introduction to Finite Mathematics.* Englewood Cliffs, N.J.: Prentice-Hall, 1966.

71. Keynes, John Maynard. *A Treatise on Probability.* London: Macmillan & Co., 1921.

72. Kissane, Barry V. "Intuitive Statistical Inference." *Australian Mathematics Teacher* 34 (December 1978): 183–89.

73. Klitz, Ralph H., Jr., and Joseph F. Hofmeister. "Statistics in the Middle School." *Arithmetic Teacher* 26 (February 1979): 35–36, 59.

74. Lake, F., and J. Newmark. *Mathematics as a Second Language.* Reading, Mass.: Addison-Wesley Publishing Co., 1977.

75. "Landon, 1,293,669; Roosevelt, 972,897." *Literary Digest,* 11 October 1936, pp. 5–6.

76. Leffin, W. *A Study of Three Concepts of Probability Possessed by Children in the Fourth, Fifth, Sixth, and Seventh Grades.* ERIC No. ED 070 567. September 1971.

77. Lucas, James, and Evelyn Neufeld. *Picture Graphs: A Picture Book of Inventory and Graphing for Elementary Grade Children.* San Leandro, Calif.: Educational Science Consultants, 1973.

78. Mathematics Resource Project. *Statistics and Information Organization.* Palo Alto, Calif.: Creative Publications, 1977.

79. McGill, R., J. W. Tukey, and W. A. Larsen. "Variation of Box Plots." *American Statistician* 32 (1978): 12–16.

80. McNeil, D. D. *Interactive Data Analysis.* New York: Wiley-Interscience, 1977.

81. Montagu, Ashley. *Human Heredity.* 2d rev. ed. New York: World Publishing Co., 1963.

82. Moody, Paul Amos. *Genetics of Man.* 2d ed. New York: W. W. Norton & Co., 1975.

83. Mosteller, Frederick. *Fifty Challenging Problems in Probability with Solutions.* Reading, Mass.: Addison-Wesley Publishing Co., 1962.

84. Mosteller, Frederick, Robert E. K. Rourke, and George B. Thomas, Jr. *Probability with Statistical Applications.* Reading, Mass.: Addison-Wesley Publishing Co., 1970.

85. Mosteller, Frederick, William H. Kruskal, Richard F. Link, Richard S. Pieters, and Gerald R. Rising, eds. *Statistics by Example.* Reading, Mass.: Addison-Wesley Publishing Co., 1973.

86. Mosteller, Frederick, and R. E. K. Rourke. *Sturdy Statistics.* Reading, Mass.: Addison-Wesley Publishing Co., 1973.

87. National Advisory Committee on Mathematics Education (NACOME). *Overview and Analysis of School Mathematics, Grades K-12.* Washington, D.C.: Conference Board of the Mathematical Sciences, 1975. Available from the National Council of Teachers of Mathematics.

88. National Council of Teachers of Mathematics. *Collecting, Organizing, and Interpreting Data.* Topics in Mathematics for Elementary School Teachers, no. 16. Washington, D.C.: The Council, 1969.

89. _____. *The Growth of Mathematical Ideas, Grades K-12.* Twenty-fourth Yearbook. Washington, D.C.: The Council, 1959.

90. Nuffield Mathematics Project. *Pictorial Representation.* New York: John Wiley & Sons, 1967.

91. _____. *Probability and Statistics.* London: Newgate Press, 1969.

92. Ockenga, Earl, and Joan Duea. "Ideas." *Arithmetic Teacher* 23 (December 1976): 623-30.

93. _____. "Ideas." *Arithmetic Teacher* 25 (May 1978): 28-32.

94. Ore, Oystein. "Pascal and the Invention of Probability Theory." *American Mathematical Monthly* 67 (May 1960): 409.

95. Pagni, David L. "Applications in School Mathematics: Human Variability." In *Applications in School Mathematics,* 1979 Yearbook of the National Council of Teachers of Mathematics, edited by Sidney Sharron. Reston, Va.: The Council, 1979.

96. Pereira-Mendoza, Lionel. "Graphing and Prediction in the Elementary School." *Arithmetic Teacher* 24 (February 1977): 112-13.

97. Piechowiak, Ann B., and Myra B. Cook. *Complete Guide to the Elementary Learning Center,* pp. 187-90. West Nyack, N.Y.: Parker Publishing Co., 1976.

98. Pincus, Morris, and Frances Morgenstern. "Graphs in the Primary Grades." *Arithmetic Teacher* 17 (October 1970): 499-501.

99. Råde, L. *Probability and Simulation, an Experimental Test for Secondary Schools.* [In Swedish.] Lund, Sweden: Liber Laeromedel, 1977.

100. _____. *Take a Chance with Your Calculator: Probability Problems for Programmable Calculators.* Forest Grove, Oreg.: Dilithium Press, 1977.

101. Engel, Arthur. "Introduction to Computer Programming," chap. 16. In *Elements of Mathematics, Book 0, Intuitive Background.* St. Louis: CEMREL, 1973.

102. Rapoport, A. *Two-Person Game Theory.* Ann Arbor, Mich.: University of Michigan Press, 1966.

103. Reinhardt, Howard E., and Don O. Loftsgaarden. "Using Simulation to Resolve Probability Paradoxes." *International Journal of Mathematics in Science and Technology* 10 (1979): 241-50.

104. Roberts, F. S. *Discrete Mathematical Models.* Englewood Cliffs, N.J.: Prentice-Hall, 1976.

105. Romberg, T. A., J. G. Harvey, J. M. Moser, and M. E. Montgomery. *Developing Mathematical Processes.* Chicago: Rand McNally, 1974, 1975, 1976.

106. Royal Statistical Society Committee on the Teaching of Statistics in Schools. "Interim Report." *Journal of the Royal Statistical Society,* Series A, 131 (1968): 478-99.

107. San Diego County Department of Education. *Mathematics for Early Childhood,* pp. 75–79. San Diego, Calif.: The Department, 1974.

108. Schlesinger, Arthur M., Jr., ed. *History of American Presidential Elections 1789–1968.* Vol. 3. New York: Chelsea House, 1971.

109. Schroeder, Lee L. "Buffon's Needle Problem: An Exciting Application of Many Mathematical Concepts." *Mathematics Teacher* 67 (February 1974): 183–86.

110. Scott, Louise Binder, and Jewell Garner. *Mathematical Experiences for Young Children: A Resource Book for Kindergarten and Primary Teachers,* pp. 152–55. San Francisco: McGraw-Hill Book Co., 1978.

111. Secondary School Mathematics Curriculum Improvement Study. *Course 5.* New York: Columbia University, Teachers College Press, 1972.

112. Shaughnessy, J. M. "Misconceptions of Probability: An Experiment with a Small-Group, Activity-based Model Building Approach to Introductory Probability." *Educational Studies in Mathematics* 8 (1977): 295–316.

113. Shubik, M. *Games for Society, Business and War.* New York: Elsevier, 1975.

114. ———. *The Uses and Methods of Gaming.* New York: Elsevier, 1975.

115. Shulte, Albert P. "A Case for Statistics." *Arithmetic Teacher* 26 (February 1979): 24.

116. Shulte, Albert, and Stuart Choate. *What Are My Chances?* Books A and B. Palo Alto, Calif.: Creative Publications, 1977.

117. Sigas, Suzanne. "Making and Using Graphs." *Instructor* 86 (November 1976): 98–100.

118. Simon, Julian L., David T. Atkinson, and Carolyn Shevokas. "Probability and Statistics: Experimental Results of a Radically Different Teaching Method." *American Mathematical Monthly* 83 (November 1976): 733–39.

119. Simon, Julian L., and Allen Holmes. "A New Way to Teach Probability Statistics." *Mathematics Teacher* 62 (April 1969): 283–88.

120. Singleton, R. R., and W. F. Tyndall. *Games and Programs—Mathematics for Modeling,* pp. 70–82. San Francisco: W. H. Freeman & Co., 1974.

121. Smith, David Eugene. *History of Mathematics.* Vol. 2. New York: Dover Publications, 1958.

122. ———. *A Source Book in Mathematics.* New York: McGraw-Hill Book Co., 1929.

123. Smith, D. G. "Apple II High-Resolution Graphics." *Kilobaud Microcomputing,* September 1979, pp. 104–6.

124. Snell, J. L. *Introduction to Probability Theory with Computing.* Englewood Cliffs, N.J.: Prentice-Hall, 1975.

125. Sobol, I. M. *The Monte Carlo Method.* Chicago: University of Chicago Press, 1974.

126. Sokal, Robert R., and F. James Rohlf. *Biometry: The Principles and Practice of Statistics in Biological Research.* San Francisco: W. H. Freeman & Co., 1969.

127. Sootin, Harry. *Gregor Mendel: Father of the Science of Genetics.* New York: Vanguard Press, 1959.

128. Souviney, Randall J. "Quantifying Chance." *Arithmetic Teacher* 25 (December 1977): 24–26.

129. Stockton, Frank R. *A Storyteller's Pack: A Frank R. Stockton Reader.* New York: Scribner's, 1968.

130. Stuart, F. *FORTRAN Programming,* pp. 294–95. New York: John Wiley & Sons, 1969.

131. Tanur, Judith M., Frederick Mosteller, William H. Kruskal, Richard F. Link, Richard S. Pieters, and Gerald R. Rising, eds. *Statistics: A Guide to the Unknown.* San Francisco: Holden-Day, 1972.

132. Thompson, M. *Probability and Statistics.* Mathematics Methods Project, Indiana University. Reading, Mass.: Addison-Wesley Publishing Co., 1976.

133. Travers, Kenneth J., LeRoy C. Dalton, and Vincent F. Brunner. *Using Advanced Algebra.* River Forest, Ill.: Laidlaw Bros., 1975.

134. "Trial by Mathematics." *Time,* 26 April 1968, p. 41.

135. "Trying to Measure Hardship." *Time,* 12 February 1979, p. 48.

136. Tukey, J. W. *Exploratory Data Analysis,* pp. 2–55. Reading, Mass.: Addison-Wesley Publishing Co., 1977.

137. Tversky, A., and D. Kahneman. "Availability: A Heuristic for Judging Frequency and Probability." *Cognitive Psychology* 5 (1973): 207–32.

138. _____. "Belief in the Law of Small Numbers." *Psychological Bulletin* 76 (1971): 105–10.

139. Ulam, Stanislaw M. *Adventures of a Mathematician.* New York: Scribner's, 1976.

140. _____. "Computers." In *Mathematical Thinking in Behavioral Sciences,* pp. 163–73. San Francisco: W. H. Freeman & Co., 1968.

141. "What Went Wrong with the Polls?" *Literary Digest,* 14 November 1936, pp. 7–8.

142. Wilson, Bryan. "'The First Shall Be Last." *Teaching Statistics* 1 (May 1979): 53–56.

Projects

Compiled by
Ralph H. Klitz, Jr.

Automobile Makes

Determine the most popular make of car or truck in your community. Is it a Ford, Chevy, Plymouth, or some other make? Are the subcompact cars, the four-wheel drive vehicles, or pickup trucks more popular? Find out by taking a survey at a busy intersection. Remember that the time of day and the day of the week you choose for your survey may influence your data. It is also interesting to check the new car registration bureau in your area to determine the current frequency of registration for certain makes of vehicles.

Automobile Traffic Patterns

Determine the number of vehicles that pass a particular point within various time segments during the day. From these data, try to project the number of vehicles that might pass the point during a twenty-four-hour period, during daylight hours, or on a Saturday or Sunday. After this project is completed, invite a traffic engineer into the classroom to discuss traffic patterns and traffic flow at an intersection or along a highway. The engineer might even have information on the intersection that the class has surveyed. Be sure to inquire how a traffic survey is conducted and what equipment the survey crew uses.

Automobile Colors

Ask students to record the colors of the automobiles that pass through a busy intersection in your community. Have them use a bar graph to present the results of their data. Alert students to the fact that manufacturers can provide reference data on the most popular colors of automobiles and light-duty trucks they produce.

The M&Ms Experiment

Ask each student, as a homework assignment, to bring a bag of M&Ms to class. Any size of bag will do, but larger bags will yield more data. Have all students count out their own candy in paper plates, recording the number of candies of each color and also the total number in each bag. Once the totals for each color and the percent each color is of the total are computed, the class is ready to analyze the results. What color occurs most often? Next often? Are any colors missing from some bags that occur in other bags? Why? It is also worth noting whether the number of M&Ms varies in different bags of the same size. How many do you think the manufacturer intended to put in each bag?

Finally, the students may eat the data!

Approximating Pi

Consider the technique of approximating pi using the Monte Carlo method. If a circle is inscribed in a square, the ratio of the area of the circle to the area of the square will be $\pi/4$. For the sake of convenience, consider the square and an inscribed circle to be in the first quadrant. If the square has sides of length 2, the circle will have a radius of 1. Using a computer, generate random ordered pairs from 0 to 2. Test each pair generated to determine if it satisfies the inequality $(x - 1)^2 + (y - 1)^2 \le 1$. It will be necessary to count those ordered pairs that do fall on or within the circle and compare them to the total number of ordered pairs generated. This ratio will of course be $\pi/4$, which if multiplied by 4 yields π. Generate as many ordered pairs as you like to improve the approximation for pi. There will, however, be an upper limit on the accuracy of the approximation based on the computer system used.

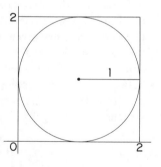

Attendance at School

Study the attendance figures at your school for a month or more. Make a histogram of the number of students absent each day during that period of time. On what day of the week do most students seem to be absent? Are you surprised? Can you determine the factors that contribute to extremes in the absentee rate, either when it is very high or very low? Does the weather have anything to do with the results? How about religious holidays or other factors? What is the average number of students absent each week? Each month?

School Tardiness

Can you determine whether there is any relationship between the distance a student lives from school and the number of times that student arrives late? Is it possible that those who live close to school have a higher probability of being late? Collect your data from the attendance office, and you are ready to go to work on the project.

Tossing Chips to Make a Histogram

Place a large piece of posterboard flat on the classroom floor with one edge against a wall. Have the students toss plastic chips onto the posterboard from a given distance, aiming them to land as close to the wall as possible. Students get a predetermined number of tosses; each time they retrieve their chip, they should mark the place where it landed with an X and their initials. If several students will be tossing at one time, disputes can be avoided by identifying the chips in some way—perhaps with initials penciled on an attached square of gummed paper.

When the activity is completed, the students can measure the distances from the wall to the Xs, find their individual scores (the lower score being the best), and determine the class average. This information can then be plotted as a histogram with students' names on the horizontal axis and their score averages on the vertical axis.

Word Hunt

Ask students to bring all their textbooks to mathematics class one day (assuming they all use the same science, history, and English book, for instance). From a given book, pick a page number at random, and ask students to count and record the number of words that occur before the first *the* on that page. Consider, in addition, having them count the number of *the*s on the page. Do the same with each textbook, keeping a record for each. What observations can be made with regard to these data? Is there a difference between the frequency of the word *the* in their mathematics book and in their English book? Repeat this experiment for other words (for example, *an, segment, irrational, direct,* and *radical*). What can be said, on the basis of the data, about the different writing styles?

Consider also the possibility of studying the length of words in a particular text.

Estimating Distances

With two pieces of masking tape, mark off a distance of three meters on the classroom floor or wall. Ask students to estimate the distance; record all estimates, and display them with a histogram. Find the mean, median, mode, and standard deviation. What shape does the graph have? Which student most closely estimated the distance? Can the class as a group judge distance well? Try the experiment again to see if the students' powers of estimation have improved.

This project can be done outside in an open area with a longer measured distance. Does the location of the project influence the outcome?

Tossing Chips

Make a line on the classroom floor with a piece of narrow tape or string. Have the students toss pennies or plastic chips, trying to have them land on the line. Keep a record of the number of chips that land on the line and the distance the remaining ones are from the line. Where is the chip most likely to land—on or off the line? What is the mean distance from the line for those that are off? Are there any "outliers"? What happens if the activity is conducted with the line placed along the base of the wall?

Statistics Bulletin Board

Having a bulletin board to display material relating to statistics will do much to increase your students' awareness of statistics. Ask students to bring in clippings from newspapers and magazines that give statistical data. Give them an opportunity to explain their article to the rest of the class before posting it. In addition, each time the class or a group of students completes a project, one of them should be able to summarize the results for the class. Then the project can be placed on the statistics bulletin board.

Outside Reading

Ask students to choose a topic that interests them from the book *Statistics: A Guide to the Unknown* [44] and present an oral report to the class. You could also have them read Huff's book *How to Lie with Statistics* [58], any issue of *Consumer Reports* [26], or the daily newspaper as good alternative sources of information for the report.

Advertising

Ask each student to choose an advertisement from a magazine and examine the advertiser's claims. The advertisements of automobile, television, and toothpaste manufacturers are favorites of students. If there is some doubt concerning the validity of an advertiser's claim, have the student write the advertiser for information to back up the claim. In the United States, advertising claims must be supported with data, and these data are available on request.

Consider this a long-term project; it may take several weeks or months to complete.

Using the Computer to Illustrate
the Central Limit Theorem

The computer can be used to generate data and frequency histograms to illustrate the central limit theorem. Using the computer to simulate, select samples of size n with replacement from a box containing balls numbered 0 to 9. The mean and standard deviation of each sample of size n = 6, 12, 18, 24 are recorded. Obtain 100 such means of each sample size as well as the frequency distribution.

Using the Computer to Toss Coins

Give students an opportunity to toss coins and record the number of heads and tails from their experiment; then simulate the project on the computer. Using the random number generator function (on most systems, RND(0)), which will produce values $0 < r < 1$, assign H to $0 < r < 0.5$ and T to values $0.5 < r < 1$. Print H, T for as many tosses as desired.

Random Single Digits

Use the computer to generate random single digits. After a short run, stop and ask what the probability is for obtaining a 4; then determine the probability for two 4s in a row. Plot the output of the computer to determine whether the digits are random. What will the histogram look like? Use an appropriate test to determine if the digits are, in fact, random.

First Names

What is the most common first name for girls in your school? For boys? Consider sampling a class, a grade, or even the entire school to find out. Make a histogram for each grade level, separating boys and girls. Study the histogram carefully and rank the first names accordingly.

A related project to consider would involve checking the local recorder's office for the first names given to newborn infants during the past few months or the past year. Make a histogram and determine the most popular names for that period. Are the high points on this histogram different from those on the ones drawn earlier for the school? Do the first names parents most commonly give their children change over a period of ten years?

Safe Driving Record

Survey the entire junior and senior classes to determine the number of automobile accidents in which each student driver has been involved. Find the mean for each class and make a histogram. How do the juniors' records compare with those of the seniors? Compare your results with records available from your state department of motor vehicles. How does your school measure up?

How Long Is a Cubit?

Measure in centimeters each person's forearm from the elbow to the tip of the middle finger. Find the mean. Perform the same experiment for two more groups of students—one older and one younger than your own. Find the mean for each of these groups. Draw a histogram to indicate the data and establish a mean value to represent the students in your school. How does this compare to the established value for a cubit?

What is the history of this unit of measure?

Vertical and Horizontal Distances

On the chalkboard draw a horizontal line one meter long. Near it, draw a one-meter vertical line. Now survey the students and adults in your school, asking which line appears to be longer. Have each participant estimate the length of both lines. You know the lines are the same. How do others perceive the horizontal versus the vertical line? For which are the estimates more accurate? Make a histogram of the results and draw possible conclusions from the data you have collected.

Finding the Area of a Lake

Use the Monte Carlo method to find the surface area of a small lake or pond near your school. Use rope to mark off an appropriate coordinate system, and choose teams of two students to begin the work of locating the points. With 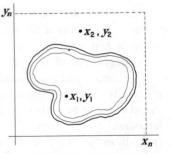 either a table of random numbers or a programmable calculator to generate random ordered pairs, begin to locate the ordered pairs by sighting them. Determine whether each point lies on the surface of the lake or the land surrounding it. The ratio of the number of points situated on the lake to the total number of points will be equivalent to the ratio of the surface area of the lake (unknown) to the area of the measurable rectangle. Plot as many points as possible. A number of teams can locate a large number of points in a relatively short period of time.

Random Numbers

Using your computer or programmable calculator, write a program to generate random digits (0–9). After generating 100 or more numbers, make a histogram illustrating the number of times each digit occurs. What shape does the histogram have? If possible, program your computer to plot a histogram after generating several thousand random digits. What is the mean and standard deviation of this data set?

Pi

The fraction 355/113 is a rather respectable approximation for pi. Write a computer program for the division algorithm so that you can perform long division to any desired number of places. Study the distribution of the digits of the number 355/113 for the first 100 digits. Do you think the digits are randomly distributed? (You must look at 112 digits before repetition begins on the 113th digit.) Find other rational numbers with long cycles of repetition and study their distribution of digits. What can you say about 1/4 or 2/9?

Retail Space versus Population

Determine the total area of retail space in your community and the total amount of money spent last year in retail stores. Use these figures to find a realistic estimate of revenue per square meter. Check the amount of new retail space planned and the projected population to verify the advisability of the expansion.

Maze

Construct a maze from balsa stripping. One at a time, have blindfolded persons move a finger through the maze and memorize it. Check the time each individual takes to learn the maze against such variables as sex, age, or the time of day.

Attitudes

Use the chi-square test to determine whether the male population at your school has a different attitude toward coeducational physical education and competitive sports from the female population.

Reaction Times

Have a student who is interested in constructing electronic devices build a unit to measure reaction time. Use the device in different experiments where data on reaction time can be collected and analyzed. For example, in your classroom compare the boys' reaction time to a given stimulus to that of the girls; compare reaction time of students with driver's licenses to that of students without driver's licenses.

Probabilistic Number Theory

Choose two positive integers at random in the interval from 1 to N. Find the greatest common divisor (GCD) of the numbers using the Euclidean algorithm. Study by simulation the random variation of the number of divisions to obtain the GCD. Estimate by simulation the probability that two integers chosen at random in the interval from 1 to N are relatively prime.

Noncollinear Points

Given three noncollinear points in a plane, employ a computer search to locate the point that minimizes the sum of the distances to the other points. Repeat to determine if any conjecture about the location of the points can be established.

The Family

Survey a class in your school to determine the size of each child's family and the number of boys and girls in each family. Compare these results to the national figures on the mean number of children per family and on the ratio of boys to girls.

Tossing a Die

Using a programmable calculator or a computer, simulate the tossing of a die. Run a number of simulations and record the length (number of tosses) necessary to see all the numbers on the die. What does the average length of this run seem to be?

Rank Order Coefficient

Consider using the rank order coefficient to determine whether a relationship exists between the number of runs scored by a baseball team and the team's ranking in its league.

Efficiency of Sorting on Your Computer

Compare, on your computer system, the efficiency of two sorting algorithms, such as the bubble sort and the Shell-Metzner sort [120]. Display the results of studying the execution time, the number of paired comparisons, and the number of pair switches. This information may prove useful later in handling large sets of data.